RAMBLING MAN

MY LIFE ON THE ROAD

Also by Billy Connolly

Tall Tales and Wee Stories
Windswept & Interesting

BILLY CONNOLLY

RAMBLING MAN
MY LIFE ON THE ROAD

TWO
ROADS

First published in Great Britain in 2023 by Two Roads
An imprint of John Murray Press

1

A CIP catalogue record for this title is available from the British Library

Hardback ISBN 978 1 399 80257 4
Trade Paperback ISBN 978 1 399 80258 1
ebook ISBN 978 1 399 80259 8

Typeset in Celeste by Palimpsest Book Production Ltd, Falkirk, Stirlingshire

Printed and bound in Great Britain by Clays Ltd, Elcograf S.p.A.

John Murray policy is to use papers that are natural, renewable and
recyclable products and made from wood grown in sustainable forests.
The logging and manufacturing processes are expected to conform
to the environmental regulations of the country of origin.

Two Roads
Carmelite House
50 Victoria Embankment
London EC4Y 0DZ

www.tworoadsbooks.com

John Murray Press, part of Hodder & Stoughton Limited
An Hachette UK company

To Pamela,
who has tolerated my ramblings and restlessness
for over forty years, jam sandwich at the ready . . .

CONTENTS

RAMBLING MAN
MY LIFE ON THE ROAD

PROLOGUE

SETTING OUT

WHEN I WAS a wee boy, I felt like an outsider. I didn't fit in anywhere. On Sundays, when my father would take me and my sister to the markets, we'd walk a few miles back home, and on the way I used to go into litter bins by the side of bus stops and fish out the cigarette packets. I would take the silver paper out and roll it together until I had a big ball of it. I was in the Cub Scouts, so I took it to them hoping to get my charity badge by selling my ball as scrap and giving the money to charity. But I never got that badge. The women in charge said: 'You can't just make one ball and expect to get your badge. You have to save up a lot of balls.' I had fallen short again. I thought: 'I'll do something else, then.' I decided to ask people for their broken jewellery so I could make something of it, but I never got round to that either. I was a failure at most things I tried.

My whole life has been like that. As a boy, I could never follow the established rules and paths that could lead to the kind of success other boys achieved. I never joined a cycling club. I never joined a hiking club or a camping club. I *knew* I wouldn't fit in. I knew that when I got among other members, I wouldn't do what you're supposed to do to get the prizes. I'd be a flop.

The one thing that made me feel less of an outsider was to be alone on the road. I'd leave the house and just start walking. I would go along the main street and see where it led me. I'd see the canal winding ahead, so I'd follow that for a while, wishing I had my fishing rod. I'd pass the Singer sewing machine factory and other places I'd seen before. I felt perfectly comfortable just walking by myself. I was on my own at last, and I felt enormously relieved. I realised that if I fitted in anywhere in the world, it was here, just rambling along to nowhere in particular. I walked in streets, graveyards and parks. I've always liked graveyards. I'd read the headstones and wonder who those people were.

Hairy Mary used to say 'Hello' to me. She was an old lady I'd often see on the street. When I was coming to her corner, she would say: 'Weeelll?' and stand on her hands. Her skirt would fall over her head so you could see her old knickers, but her hat always stayed on. She was a wild woman of the streets, and I was proud that she knew me. She'd say, 'Hello son!' I'd say 'Hello! How you doing, Mary?' She'd say: 'I'm not Mary. They just call me Hairy Mary.' I'd say, 'Then what's your real name?' She'd say, 'None of your fucking business.' She was always angry, and I loved her for it. She wandered the streets, same as me. She never said: 'Aren't you a bit young to be wandering about on your own?' like other adults did. I'd say: 'Where do you live?' 'None of your business.' 'Okay.' But it was obvious she lived on the street. She was always complaining about people who'd moved her on. 'I don't like that janitor' . . .

I never worried about getting lost. Nobody's ever really lost. You just walk until you become unlost. I walked in Kelvingrove Park, and in Victoria Park, where miniature yachts were sailing on the pond. Beautiful things. Huge. They'd set them out to sail, then run round and turn them onto another tack. An owl lived in Victoria Park, in a cave. I'd peer in and look at it and it would stare back. And there were petrified trees from the Stone Age. I would sit there among them, and I felt calm. I never wanted to go home, but when it got near dinner time I'd start heading back, because if I didn't turn up at dinner time they'd realise I was missing.

———

Then one day I came across a picture that inspired me more than any of the images of heroes or saints or movie stars I'd seen. It was a colourful drawing of a cottage somewhere in the countryside, with a lovely garden and roses growing all round the doorway, and there was a woman standing at the door talking to a tramp. He was wearing a well-worn hat and a coat with rope tied round the waist instead of a belt. He was munching a jam sandwich that the woman had obviously given to him and he looked so happy, thoroughly content and pleased with his life. It seemed like a legitimate, grown-up progression from my boyhood walking habit. I remember thinking, 'I want to be like him when I'm older . . .'

That feeling has never left me, of wanting to be free and wandering wherever I pleased. I've always loved the idea of taking off and living a life where I had no responsibilities, where I could be free to go anywhere in the world, whenever I pleased. When I was in my twenties, I thought it was possible that I could live off my wits and just work when I needed to. I could play music and tell stories – a bit like a medieval troubadour. I imagined that I would meet plenty of people who would be kind to me, give me food and put me up for the night. The call of the road was so strong in me that, even after I was married to my wife Pamela, I once proposed that I should become a full-time hobo – I'd stride out onto the open road and just turn up at home whenever I needed money. I dearly wish that conversation had gone a bit better than it did.

There are many different types of migratory folk and all kinds of names for us too, including 'hobos', 'drifters', 'vagrants' and 'bums', but you can't use most of those terms any more. In any case, it doesn't matter which wandering subset you belong to – whether you have lodgings or not, whether you work, how far you travel, whether you ever leave home or just dream about it – I think any man or woman who's got a faraway look in their eyes is a Rambling Man.

I've known many Rambling Men throughout my life. Some of them travelled, either temporarily or permanently, and some were simply Rambling Men at heart. Those who travelled did so not because they didn't have a choice, but because in a more settled environment they felt like an outsider. Such people have a deep and painful sense that they don't belong. They're in a place where they don't fit, so they wander off. They meet other people who've wandered, and they get great comfort from the fact that someone else feels the same way.

Rambling Men know each other when they meet. They're the same. You can go into a café, and you look over at a table where a particular person is sitting and you go, 'How are you doing?' You can recognise the fellow Rambling Man by certain signs: his denim, demeanour, attitude. A beard used to be a dead giveaway. When I was with my first wife, we'd be walking along Sauchiehall Street in Glasgow and there'd be another guy with a beard approaching, and she'd say: 'Do you know him?' I'd go, 'Hello, Alec.' 'Hello, Billy.' People with beards knew each other. Not like now.

A true Rambling Man is someone who doesn't owe anything to anybody. He doesn't owe them money; he doesn't owe them his time or his talent. He's a free man. Free to do what he wishes, when he wishes. He may be settled in one place but, in his heart, he doesn't see it as permanent. A Rambling Man doesn't want to be tied down to anything for ever, either one job, or one way to travel, or one place; he lives with his desire to move on. He walks, he bums lifts, he buys tickets – he travels in any way he pleases. He lives on his wits, making money by doing whatever he's good at. He's talented at practical things, so he can repair houses or help on a farm – do various odds and ends to make a living. He might get a job tidying up people's gardens, sweeping their premises or walking their dogs. Anything. And some jobs are fucking hard. Traditionally, harvest was a time when there were plenty of jobs for rambling folk, like picking potatoes or apples. A Rambling Man might have exceptional skills he can earn money from. Welding was one; he could make metal signs for people's houses. Or he might be an artist or musician who can sing songs, recite poetry, paint or play the guitar. People put him up because they like him, and he helps them out in all kinds of ways.

A Rambling Man is often self-educated. The fellow Rambling Men I came across were generally well-read. In the early days, they tended to read poetry by the likes of Allen Ginsberg, but I always preferred the more traditional poems from my schooldays. I liked the simple stuff . . . and Robert Burns, who is anything but simple. You'd memorise a line or two from a particular poem you loved and recite it to a fellow Rambling Man, and even if he didn't know the poem he'd tell you what he thought of it, then recite one of his own. We all had writing heroes and we shared them with each other. Everybody I met in the sixties and seventies had read *On the Road* by American writer Jack Kerouac, who'd wandered across America with various friends. Hemingway was popular too – I read him a lot, as well as John Steinbeck's short novels and stories. My favourite was *The Pastures of Heaven*. At the heart of all these books was a sense that there was something more to life in a place somewhere beyond where you were; that adventure was on the horizon, you just had to stride out towards it.

When I'm not on the road I don't long for it at first, but then this feeling begins to creep back and I miss it a lot, because rambling has always been my natural function in life. I think most of the Rambling Men I meet are escaping from something. Sometimes it's just avoiding boredom. The good job syndrome. People say: 'You've got a good job so why be restless?' It's because you knew there was a whole world out there – countries to visit, things to see and adventures to have – that wouldn't come so easily if you stayed in one place. You just had to put yourself out there and they'd come your way – that was the belief. Exposing yourself to new places and people, and being part of it all, would eternally make them a part of you.

———

A woman can be a Rambling Man. One of the best Rambling Men I ever met was a woman in Texas who lived in a taxi. She was seventy-one and she was brilliant. She *looked* brilliant too. High cheekbones and a very positive attitude. I think it's your attitude to life that determines the way you look. If you're a weary willie and moan all the time and think you've been dealt a bad hand, then you'll probably look trippy-faced. But if you're open to nice things and wish people well – it shows in your face. If you're optimistic, people want to be with you. If you're not, they don't want to come near you. This taxi-dwelling woman was extremely appealing. Women who are fiercely independent and are living on the edge always have a great look about them. I want to spend time with them. I want them to be my friend. I've met a lot of women like that who are great Rambling Men. Buy you a beer. 'Don't take shit from no one.'

———

To be a Rambling Man, you don't have to live your life on the road. It's more a state of mind. An ideal. A Rambling Man can seem like a perfectly regular person – someone who goes to work every morning, has a family and pays his taxes. He might look conservative and behave in a normal fashion but, deep down, he has the *spirit* of a Rambling Man. He harbours *longing*. Maybe he has a half-built Harley-Davidson motorbike in his garage and listens to Bob Dylan while he's working on it. He has *hope* inside him – the hope of travel and freedom and of being who he really wants to be; that's the difference between him and others around him. It's the opposite of living miserably. Some people stay trapped, paralysed in private despair.

I've been a Rambling Man since I was in short trousers, but a Rambling Man can be any age. In fact, having the sensibility of a Rambling Man helps to keep you young. Once I came into my seventies, I noticed it wasn't considered that old any more – and I'm not just saying that because I'm eighty. The number of times I hear people say, 'God! He died? But he was so young! Only seventy-four!' When I was a boy, fifty was old and sixty was ancient. But my fifties were great. I heard about Jeanne Calment, a Frenchwoman who lived to be 122. She was the oldest person who ever lived. She said: 'I think God forgot me.' She claimed to have met the painter Vincent van Gogh, whom she thought was a desperately ugly man who smelled of alcohol. Apparently, she began fencing when she was eighty-five and was still riding her bike when she reached 100. 'I only have one wrinkle,' she said, 'and I'm sitting on it.'

A Rambling Man prefers to make a journey for pure pleasure, although sometimes it must be made out of necessity. I've had some wonderful trips in my lifetime, but my most recent wasn't one of them. My back has been giving me gyp, so this one time I had to wake in the middle of the night and let my wife drive me to Miami for an early appointment at a spinal place. Four hours later, I was snoozing loudly in the waiting room when the nurse called me in for a poke, a prod and a series of impertinent questions. Then the doctor came in. 'Good morning, Mr Connolly. It's a real honour to meet you! I'm going to shock you with electricity then stick you with needles . . . how does that sound?'

Fucking great.

The drive back home from Miami was the same route in reverse, but it was decidedly more pleasant. There was the same pearlescent sky, jade sea, series of palm-covered islands linked by storm-lashed bridges. The helicopters were still hovering over the Seven Mile Bridge, dangling the brave buggers whose job it is to repair the power lines. The same tiki hut signs welcomed visitors for grouper sandwiches and cracked conch, washed down with 'bottomless' margaritas . . . but this time, with all the pricking and prodding behind me and the prospect of a nice cup of tea in my own easy chair, it was nothing short of delightful. The reason for your journey can change how it feels, but it's all about your state of mind. Many of my past journeys have been made because of work necessity – just a race in the dark to get from one town to the next for my concerts. I wasn't able to enjoy even the most brilliant landscapes when I knew I'd soon be facing audiences with high expectations. But I've also made many journeys that were just for my own pleasure, and a few that were filmed for television. In this book, I'm going to focus on the journeys I've truly enjoyed – even if a film crew came along for the ride.

———

I'm a showbiz personality. I read it in the paper. I'll be on *Survivor* soon, with a fucking DJ that nobody knows. It's great being in showbusiness. When you're a Rambling Man who also happens to be a showbiz personality you get to do extraordinary things, like be one of the few people to climb to the top of the Sydney Opera House, stand on a sacred Māori promontory in New Zealand that is usually forbidden territory for most people, or even drive an Amish man's buggy. They were one-offs, but I know I'm incredibly lucky to have done those things.

If people recognise you from TV or film, they react to you strangely. They want to show you things. They want to show you the *best* of things. I once went to play a gig in HM Prison Peterhead. When I arrived, I met the warden, and we had a pie and a cup of tea. They made their own pies there, which were delicious. Then he took me into the kitchen, where they were making dinner. A man was stirring the soup in a huge vat with an oar. He said, 'How ya doing, Big Yin?' I said 'I'm fine. How are you?' 'Ah, I'm all right. Life is good.' He told me he was from the Scottish seaside town of Helensburgh. I said: 'What did you do to get in here?' He said: 'I shot a cunt.' I said: 'Did he deserve it?' He said 'Yeah, he certainly did.' It went on like that. The prisoners were nice to me. I'm a boy from Glasgow, so they could relate to me. They brought me their drawings and things they'd made with bits of string – bracelets, for example – as if I had some kind of authority, as if my opinion meant anything. It's easy to be unappreciative and not take it seriously – I've witnessed that before. But pausing during your journey and learning about people's lives is a precious thing. It's very touching when people show you what is important to them and who they are. One guy had done some oil paintings. He was very proud of them and wanted to show me, but in order to see them I had to go into his cell. Once they shut the door behind me, it was just me and him. That made it another world altogether, one in which we were equals. It was a privilege to be shown his art, which he kept stashed down the side of the furniture because the walls of the cell were covered with pictures of naked women.

———

Travelling can be very challenging, but I realised a while ago that if I was going to be one of those guys who said, 'I don't like travelling,' then I was doomed. The audience weren't going to come and sit in my house. I decided that, if I found travelling tiring or boring, or if I suffered from jetlag, I just had to find ways to beat it. I've had to learn to get myself into a mental state of either sleep or wakefulness, depending on what I need at the time. For me, meditation has been quite useful for that. It refreshes my mind. Walking helps too.

Walking is your basic, essential Rambling Man way to travel. In my ultimate fantasy of the Rambling Man, walking – and occasionally jumping into the car of a moving train – is what he does. He slings his guitar or banjo over his shoulder and strides out along the road to the next town. There he plays a few songs at a local gig and meets a beautiful woman who feeds him and puts him up. That to me is a wonderful life. A million miles better than sitting on your couch watching reality TV. People sitting in their houses watching other people sitting in a house. Get a fucking life!

A Rambling Man believes travel – of any kind – is necessary and good for you. If you don't travel, you don't know how the rest of the world lives and thinks. And you get to know people much better by speaking directly to them than by listening to others speaking for them on TV. If you travel through places like Africa or India, you'll meet people who tell you what they would like is a bicycle, or to get the hole fixed in the roof of their house. They're not impressed by all the stuff you've got. They are content, and that's an enviable thing. I've learned a lot from them. I've found that when living is hard the spiritual value is often very high, like it is in India. Their spiritual world promises: 'Take it easy. Everything will be all right eventually.' Call me old-fashioned, but in my way of thinking that's quite a step up from the threat of roasting in hellfire for touching your willy.

In India I met a man who'd stuck a spike through his face. He was a devotee of the Hindu festival of Thaipusam that commemorates the day the god Lord Murugan used his spear to vanquish an evil demon who threatened the whole of humanity. I said: 'How does that feel?' He turned that question back on me: 'How do *you* feel?' I said, 'I feel great.' He smiled: 'I feel great too.' It was normal to him. I was very impressed by him. Many other pilgrims had skewers and sharp hooks embedded in their bodies, and they claimed to feel no pain, having fasted and entered trances before the piercings. The religion is part of their being. It's not something they just do on a Sunday. Sometimes I think they're slaves to it and it's not such a great idea, but in other ways I was deeply impressed by it and by their devotion. When you're in India you see things that make your heart sing. You'll go along the road, and you'll come to a river and there'll be a man playing a flute next to a temple. You'll think: 'There's something right about this. Something they've got that we don't have.' You don't know what it is, but THEY know.

––––

People say, 'Travel broadens the mind' blah blah blah . . . well, they're right, but I think my mind was pretty wide before I ever went anywhere because I grew up reading books about travellers and explorers. A lot of my travel has related back to my childhood – and to a yearning for places that seemed wildly romantic. Tibet, for example, where I've still never been. All the books I read about the Yukon, with the dogsleds and the howling wolves, made it so much more meaningful when I finally travelled there and discovered those things for myself. See, I was never much interested in the learning offered to me at school, but my journeys have given me the best kind of education – lessons learned in a way that made them unforgettable. This is a stark contrast to my early classroom days, where a teacher would come in: 'Today we're going to learn Pythagoras' Theorem . . . Connolly! Go out and count the railings, there's a good chap.'

Travelling for its own sake as Rambling Men mainly do will always lead you to adventure or something of interest. But sometimes I do like the kind of travel that has a particular point to it. For example, I've often journeyed with the sole purpose of fly-fishing and, if I'm intent on catching a wee fishy, I don't notice much else. When I came back home from fishing near Tulum on the Yucatan Peninsula in Mexico, my wife was excited to quiz me about what I'd seen. 'Did you visit those amazing pre-Columbian Mayan ruins?' I must have looked blank. 'Billy! Please tell me you at least saw El Castillo?' 'Nope. But let me tell you about last Thursday, the day I caught the "Holy Trinity"– a tarpon, a bonefish and a permit . . .!'

I've learned that people respond to places in different ways depending on how they're wired. Some people are visually attuned, so they respond to colours and shapes and architecture – the beauty or the ugliness, however they view it. Other people are very kinaesthetic; they respond to the feeling of the earth beneath their feet or notice how it feels to brush through tree branches, sit on the grass or touch an ancient marble statue. Some people respond auditorily, so they notice the sounds of the wind, a rustle of the trees or music played on local instruments. Many people respond to the taste of a place. They remember the textures and flavours of the food they had there, and the taste that lingered in their mouth after they breathed in the air. Or it could be the smell – the scent of the soil, the hedges or a river. Me? I'm more of a visual person. I'll never forget going into Monument Valley. All those sculptural sandstone structures – that was the highlight for me. It was like every cowboy movie I've ever seen. I stood gazing at the vast landscape with a Morricone soundtrack playing in my head. But the Native American guy who showed me around Monument Valley brought me right down to earth. He said: 'You should come in September when the tarantulas are moving. They come to mate, and the ground is alive with them.' I said: 'I think I'll give that a miss.'

––––

Overall the most important thing that I learned, and frequently tell myself, about travelling is that it's vital to have the right attitude. The perfect Rambling Man mindset is this: Don't go to a place thinking you know all about it. Let it take you by surprise. It might be something as simple as a door handle or a floor mat that will tickle you. Make you go: 'Oh God!' Or it might be a wall monument to one of the builders who died during the erection of a building. You often just chance upon these things: the human side of monuments. I've walked into famous old houses where the heights of the children who grew up there are marked on the sides of the doorways. I love that. The house answers you back. It's a very good thing. And I've loved touches like a circular wall in a turret with a high window where people used to climb up and look out. It's wonderful to see structures that are created out of whimsy. It's a very good thing, whimsy. Without it everything is beige. Some offices are so square and cold. They have no soul. The people who built them should be fucked and burned.

I don't think I've ever had a bad trip. Well, apart from in the 1970s, but that's a whole other story. No, the only thing that would stop me enjoying a journey would be ordinariness. Something without any special features. Plain, I suppose. Though sometimes I do like ordinariness; simple things can be interesting and often very funny. I remember two women looking at raincoats in a shop window on a Sunday afternoon in New Zealand. One said to the other: 'The beige one would be very practical.' I wanted to shout: 'Get the YELLOW!'

A Rambling Man takes the bad along with the good. Sometimes a journey is awful, but that can be a good thing too. In my very early folk-singer days, I was with Frank Lynch, a former manager, on a plane flying to the USA and I was drunk. I was crazed and I was shouting: 'Take me back to Earth! I am an Earthling! I don't belong up here!!' The main cabin was jam-packed and people were complaining. Frank was no use; he was rolled up in a ball laughing. So, the flight attendants threw me into First Class. Well, a female flight attendant came up and said: 'Come with me. I've got a special place for you.' I spent the rest of the journey in comfort where there were hardly any passengers. Once I got up front, I decided to stop making a nuisance of myself; I just settled down and mused about what a good idea it had been to misbehave.

Awfulness can be good.

———

I've travelled by foot, bike, ship, plane, sleigh, even piggy-backed, to get all over the world; I've danced around naked in snow, wind and fire; I've played my banjo on boats and under trees; I've slept in bus stations, under bridges, on strangers' floors, and thumbed a ride through Europe and America, ever hoping some mad bugger would let me clamber into their car; I've travelled alone, with great companions and with film crews; I've thrown myself off buildings and into canyons bungee-jumping – and, in my opinion, there's no better way to live and feel alive. Being a Rambling Man was what I wanted to be, to live the way I damn well pleased and fuck the begrudgers! Along the way, I've met the weirdest and most wonderful people who walk the Earth, seen the most bizarre and the most fantastic sights – and I've rarely come across something I couldn't get a laugh at. It's just the mediocre I don't enjoy, so there won't be any of that in this book. What there will be is tales of the loveliness and the awfulness of my many journeys. I'll try to leave out anything that's plain or practical – or fucking beige . . .

1

GET YOUR KICKS . . .

———

ALTHOUGH WALKING IS the traditional means of getting around for a Rambling Man, there are times when he needs to get from A to B a bit quicker, and for those situations a bicycle or motorbike will do the job. It's just about having the means to get somewhere else fast if you're not happy in the place you're in; a simple way of staying sane. And when you're able to change the picture smartish on two wheels you can usually have a bit more fun doing it. I got my first bicycle for Christmas when I was a boy. It was a purple New Hudson racing bike, and it gave me the independence I really needed. I could escape my household and ride out into the countryside like a true Rambling Man. I had no food, no company or anything – apart from some water in a flask – but I loved it. Just being by myself where nobody could touch me or change what I was doing was brilliant. I felt like I could do or be whatever I wanted. It was all up to me, and it was a lovely feeling. I'd go up to Loch Lomond or to Helensburgh – little seaside places that were so different from where I lived in the middle of Glasgow. I loved the air up around Loch Lomond, the fresh country smells and beautiful light. Made me come over all Sir Walter Scottish:

Breathes the man, with soul so dead
Who never to himself hath said,
This is my own, my native land!

Mountains, forests, lochs . . . with a bit of luck tourism might catch on there one day, eh?

The first time I went up to Loch Lomond I was in the 141 pack of the Cub Scouts. The big camp I attended was Auchengillan. The scouts sat around a totem pole singing:

We're riding along on the crest of a wave,
And the sun is in the sky.
All of our eyes on the distant horizon,
Look out for passers-by.

————

Those Scout camps gave me a love of the countryside that has never left me, and after that first taste I often returned to Loch Lomond on my bike. I'd sit by the water to rest and make tea with little twigs. There was always a smell of tea and wood burning there. You could stop any time for a 'drum-up' using community tobacco and sugar other cyclists had left stashed in the stone walls. In my early teens, I'd just take off and go camping whenever I could. It always gave me a wee jolt to simply get up and go somewhere, get on the road, end up somewhere I hadn't been before. It was an escape, an adventure. I never really thought much about any possible danger, and my father never asked where I was going – I just went. I'd end up miles away from home, sleeping under a cheap tent – sometimes in the pouring rain – with no groundsheet and very little money for food. It never dawned on me that some people might think that was a weird thing to do.

But cycling wasn't just about escaping somewhere. I also liked to race other boys on my bike. On the edge of the housing estate where I lived, leading off the Great Western Road, were a couple of roundabouts about two or three miles apart. For me and the other boys, it was our own little racing circuit: going like the clappers down the road, then racing around the roundabouts three times and back again to the finish line. It was hard, especially in winter. In those days my cycling attire consisted of long trousers, a jumper, a thin windbreaker, fingerless gloves and a pair of Reg Harris Fallowfield cycling shoes that had Union Jacks on the tongues.

That was when cycling was simple. Thirty years later I wouldn't have dreamed of cycling in public without multicoloured spandex shorts and my Vitus Duralinox machine. When I competed in the London to Brighton charity race in the late eighties, I wore green snakeskin spandex. 'Eh, Big Yin!' One of my fellow competitors started getting on my tits. 'Those look like knickers!' 'Don't worry, pal,' I said. 'You'll only be seeing the arse.'

Cyclists are usually lovely people – when they're on their bikes. But once they get off, they join the rest of the human race. As a boy, I was drawn to the camaraderie among cyclists. I learned very early on that we needed each other. Things could go wrong – punctures, brakes or gears failing – so you had to rely on each other to keep going. If another guy had a puncture, the unwritten rule was you got off your bike and helped him, even if you were racing each other. The rationale was, there was no sense in your bashing on ahead. There was no glory saying, 'I beat you by an hour' when he was just trying to fix his puncture. You had to work together.

A similar lore exists among some types of Rambling Men, even though they are essentially loners by nature. When alone on the road they never feel lonely and do not crave company, but they do meet and chat with other people when they feel like it – sometimes with the purpose of finding a place to stay or hoping to make some cash – and they'll help each other out. One famous 'hobo' in America was Leon Ray Livingston, also known as 'A-No.1' or 'The Rambler'. In typical Rambling Man fashion, he was self-educated, and he wrote books about how to travel the 'hobo' way, and about the 'hobo code'. In America during the Great Depression when 'hobos' were everywhere, they developed a 'code' system for leaving vital messages for other travellers. These were usually secret symbols chalked, carved or etched on to signs and walls, giving directions for where to find clean drinking water, pointing to kind folk who would help you out, or giving warnings about danger or hostility in one form or another – like unsympathetic police or aggressive animals. Hobos had their own names for things too; if you ask a hobo for a 'banjo', he'll hand you a frying pan. A person who keeps the same job for a year is called a 'barnacle', while a 'California newspaper' is a blanket. It's a highly developed wee society that is secretive for very good reasons. Nowadays, there is a 'Hobo Convention' in a small town in Iowa, USA, called Britt. Every year they crown a Hobo King and Queen, and the council members hold their annual meeting to discuss matters of community importance.

My wee cycling group was far less organised. We didn't leave symbols around, but we did give each other tips about how to keep your bike rust-free or where to buy a decent bicycle pump. Over the years, bicycle gear became far more complex. For example, bicycle seats evolved into squishy pads. Some company invented an oily, rubbery substance that they trapped between two layers of leather to create seat-covers that were sort of wobbly. It felt kind of sensual, which – as you can probably imagine yourself – had the potential to lead to some new problems. I was installing one of those seat-covers on my bike but, when I read the manual, I sat up at the warning about 'penile numbness'. The manufacturers had special advice for curing it, which they described as *'discreet massage'*. 'Oh yeah?' I thought, 'Is that what they're calling it now?' I started imagining what would happen if the entire Tour de France team succumbed to penile numbness and had to resort to *'discreet massage'* during the ride. Fuck, they'd never make it over the Pyrenees. They'd be knackered – cycling into fields and going blind!

———

I graduated to motorbikes and motor trikes when I was in my twenties. The big appeal was that I could go further. I could ride to England, to Blackpool or further north in Scotland to the Isle of Skye. To escape on my motorbike was wonderful. I could get really far away to where things were quite different; to where the people lived differently and spoke differently. I met loads of other Rambling Men doing the same thing on their motor-cycles. They camped under the stars, sat round fires, and played harmonicas and guitars – the same as me. They didn't think it was peculiar at all. And neither did I. There were groups of us everywhere and it became a movement. Journalists started noticing that there was a growing number of people who lived their lives this way, and they started writing about it as a phenom-enon. I came across magazine stories about certain people from that movement and admired them, but nobody really knew quite what it was all about. We were just enjoying ourselves and living in the moment.

The feeling of being on a motorbike is second to none. You can't get that by driving a car. The wind doesn't just blow your hair. Weird fact: it makes the hippy beads round your neck fly straight out horizontally forward as you're riding along. Before I became a 'hippy' biker, I styled myself like Marlon Brando in *The Wild One* and firmly believed it made me more attractive to women. Well, at least it took the bad look off my acne. But Brando was an icon of American gang culture, not a Rambling Man. James Dean was not a Rambling Man either, but we related to the misunderstood, the loner, the outsider in him.

Although any roadworthy motorbike will probably get you where you need to go, there's no doubt about the great visual impact of being on a decent machine. At first, I had fairly ordinary second-hand motorbikes that required a lot of maintenance, but later on I had some beauties, like my purple Harley-Davidson three-wheeler. That was a killer trike. I rode it reclining, in the 'bad boy' position. It's a real 'Go fuck yourself' stance. People in beige sedans give you very old-fashioned looks. 'Don't look at him, Dorothy – next thing you'll be pregnant. He's probably got tattoos on his willy.'

———

I once had a great biker jacket made for a tour I did of New Zealand. I was making a TV show riding both North and South Islands. It's a fantastic country for riding a motorbike. The name of that tour, 'Too Old to Die Young', was written on the back in Hells Angels-style typography and with a logo of a skull. I loved it. A woman in Los Angeles created it for me. She told me she had to get permission from the Hells Angels team because they are quite particular about their style and trademarks. Anyway, I was out riding by myself one morning in Wellington wearing my jacket when I realised I was being flanked by teamsters from the Mongrel Mob – that's the largest biker gang in New Zealand, and it has quite a reputation. They surrounded me at some traffic lights and pulled me into a side street. I felt vulnerable because I was completely alone. One of them asked, 'Where'd you get your jacket?' I said, 'I got it in Los Angeles. A girl made it for me. She got permission.' 'She got permission, did she?' They were pretty aggressive. I said, 'Yeah, from the Hells Angels.' They went away and had a little conference, 'Urrrrrggggurgggg.' Then they came back and said: 'That's okay.' They backed off and let me go on my way. Afterwards, my TV crew went crazy because they didn't get any of that on film. And I didn't talk about it onstage because I wanted to get out of town in one piece.

There was a notice outside a bar I passed on the road to Wellington that said: 'This Place Welcomes Bikers.' What does that say about a bar? It says, I'm not going in. I like bikers, but there's a type of biker who goes to a biker bar that I don't want to know. Fights tend to break out, and they have good weapons – chains, padlocks. A great biker tool is a padlock with a hanky tied round the handle. The padlock goes in your back pocket and the hanky hangs out. Ready for action. A whack on the head with that thing could even be more painful than a 'Glasgow kiss', which is a sudden headbutt. Glasgow has its own distinctive headbutt that has to be experienced to be believed. You don't ever want to be on the receiving end of that, and here's a tip so you can recognise if a guy is about to headbutt you: you can tell by the way he moves his feet. He'll shift one foot in front of the other like a boxer trying to get better balance. And he'll come at you with his head to the side, then straighten up. Glasgow has its own history and culture of violence. There was a Glasgow poet called Stephen Mulrine who wrote a smashing poem called 'The Coming of the Wee Malkies' that is popular among children – even though the verb 'to Malkie' means to slash someone with a razor. People say: 'The Wee Malkies will get you' which is a threat aimed at children. That sums up Glasgow for you. 'Hard man bravado' at four years old – I'm all for it. If Glaswegians see someone with a bandage on his head, they'll say: 'Looks like he was talking when he should have been listening.' And if someone has a slash mark from a knife on his face, they'll call it a 'second prize'. On one of my many visits to a hospital emergency department, I learned that the international medical field's standard test for measuring a patient's level of consciousness after a brain injury was developed in Glasgow hospitals through the prevalence of injuries in the town after Saturday-night fights. It's called the Glasgow Trauma Scale.

The Australian town of Newcastle is a big biker town. A great group of bikers worked backstage in the theatre I played there. They were covered with elaborate ink and had braids in their hair. Their jobs were to move all the scenery on- and offstage and so on. We got on like a house on fire. They loved my Harley trike that had been in the centre spread of a magazine. During my concert I was talking about the recent Newcastle earthquake, and they all started to laugh backstage. I was kidding the audience, saying, 'You'll notice I do my whole act with my feet pointing that way. One rumble and I'm fucking off! You can hold the roof up – I'll be over there.' After the show, the biker crew told me that, during that earthquake, one of their friends who was a big, bearded biker covered in tattoos was being operated on in the hospital. They were all there in the waiting room waiting for news. When the first big rumble came, they went: 'Everybody out!' and pushed all the beds down onto the beach. So, their pal was lying in a hospital bed at the beach. When he woke up there were palm trees around him and cockatoos flying above, and he said: 'Oh! It's just like Earth!'

———

America is another great country for riding a motorcycle. The states are connected by long highways and they can get ravaged by the weather, so they are usually well maintained. When I got to ride the whole length of the famous Route 66, I found the road surfaces were spectacularly good. Route 66 is the most famous road in the world. How many other roads do people sing about? Maybe Piccadilly, Bourbon Street and Broadway. The Champs-Elysées gets a mention in a few songs . . . but Route 66 is one of those legendary places that many people long to see. It's come to represent the idea of escape, freedom and adventure, so naturally Rambling Men are drawn to it. When I was a boy becoming enthralled by rock 'n' roll, I heard: *'Get your kicks on Route 66'* and wondered what on earth that meant. I told my sister I bet it was pure magic. Thirty years later, I rode my trike for more than two thousand miles along it, from Chicago to Los Angeles, while making a TV show about the journey.

I had been to Chicago a few times in the past, but I'd never seen it from the water. Before starting to ride down Route 66 I gazed at the city aboard a boat on the Chicago River, from which I was able to get a panoramic view of some of the tallest buildings in the USA. I tried to imagine what it was like when it was a city of wooden structures – before the great fire of 1871. By all accounts, it was a relatively low-built frontier town – a gateway to the West. But almost everything in the city was burned down in that fire. Only one water tower survived – I managed to get a glimpse of it from my trike; it's almost hidden among the skyscrapers. After the fire, rebuilding began, and the steel and concrete fortresses began to rise.

One skyscraper that really stood out for me was the Tribune Tower, a Gothic Revival building, with stripes that reach three-quarters of the way up, and mad, castle-like turrets at the top. The owner and publisher of the *Chicago Tribune* was Colonel Robert R. McCormick. He'd visited Ypres in Belgium and found a bit of stone that the Germans had blown off the cathedral there during World War Two, which he kept as a souvenir. Then he started collecting more pieces of other buildings – really famous structures like the Notre-Dame Cathedral in Paris, Edinburgh Castle, the Berlin Wall, the Great Wall of China, even the Great Pyramid of Giza in Egypt. My favourite was a bit of limestone from 'Injun Joe's' cave – the one that Mark Twain wrote about. Exactly how McCormick and his retinue got these 'souvenirs' seems a bit dodgy to me, but I think I'll draw a discreet veil over that. He had the architects Howells & Hood incorporate all these bits and pieces into the facade of the building and, I must say, they look fantastic, lurking there in the middle of the 'windy city'. The place could have done with a wee bit of Glasgow, though. Maybe a chip off the old Duke of Wellington statue's plinth? No one's going to notice, with all the hoo-hah about the frequent capping of the statue; for more than forty years people have been climbing up to adorn the statue's head with an orange traffic cone. That's my town!

———

Beside Chicago's architecture, two elements in its past really grab me: gangsterdom and – best of all – the birth of rock 'n' roll. Chicago's gangsters were a rare breed. In the 1920s, when the era of Prohibition was in full swing, Al Capone ran his empire of illegal gambling, bootlegging and prostitution behind a fifth-floor bay window in the Lexington Hotel on South Michigan Avenue. Selling 'moonshine' was easy-peasy: once his barrels of non-alcoholic booze had been inspected by the authorities, he got his men to inject alcohol into them using big syringes, then sold the hard stuff in speakeasies, clandestine back-room bars with secret entrances that could be sealed off when the police showed up. Nice little racket, Scarface . . .

As for rock 'n' roll – it wouldn't have existed without the Chicago blues. There would have been no Led Zeppelin, no Rolling Stones or Allman Brothers; those giants became famous covering musicians like Muddy Waters, who were part of a musical scene that sprang up during the mass exodus of people leaving the southern states and migrating north. In the 1930s thousands of African American men and women jumped on trains and crowded into Chicago – and they brought their music with them. Their acoustic guitars, violins and harmonicas became electrified in the clubs and, pretty soon, the whole scene exploded. I'm immensely grateful to rock 'n' roll. At the root of it is a kind of rambling spirit – it's about freedom, adventure and going your own way. When I was a teenager, it saved my life. It came along right when I was despairing about my whole world – school, home, everything. Whenever you feel like a no-hoper, just play 'Heartbreak Hotel'. You'll know for sure you're not alone.

Well, since my baby left me
Well, I found a new place to dwell . . .

———

Rock 'n' roll was not the only music that developed in Chicago. I went to a service at the famous Quinn Chapel, home to the oldest black congregation in the USA, and heard the most brilliant, stirring, powerful gospel music. The Quinn Chapel was a place where freed slaves, escaped slaves and abolitionists gathered, and it's one of the places where gospel music really began in America.

The service was full of extraordinary spectacles, such as the preacher coming in like Jesus, carrying a giant cross on his back. My boyhood Catholic Masses were nothing like that. If they had been, I'd probably still be attending. Instead of threatening us with hellfire, a couple of verses of Lee Roy Abernathy's song 'Everybody's Gonna Have a Wonderful Time Up There' would have made all the difference. It was such a missed opportunity by those priests back then. I could have been swinging the thurible going '*Doo Wop Doo Wop*'.

———

Before I left Chicago, I went to the top of the Willis Tower – formerly the Sears Tower – which was the world's tallest building for twenty-five years from 1973. Spectacular views. I walked out onto a wee observation pod at 110 storeys up, perched my cowboy boots firmly on the glass floor and looked down. *Whoooah.* I don't usually mind heights, but this was the creepiest feeling. From this glassed-in crow's nest, I could see the start of Route 66 – the road I was about to travel 2,500 miles down, across eight states, right through the heart of America. Suitably oriented west, I put on my leathers and my chaps with the open crotch to keep my bollocks nice and ventilated. I saddled up, put on my helmet with its hidden microphone, and climbed on my trike. At the traffic lights, I chatted to a few fans, posed for some snaps – and then, eventually, I saw it: the sign that said: *'ROUTE 66 BEGIN'.* Yippee! But . . . you'd think they'd put *'ROUTE 66 STARTS HERE'*, wouldn't you? But no, it was more like a command: *'BEGIN!'*

As instructed, I began. I was off in the direction of Pontiac – my first main stop on America's most famous highway – with the cameras pointed at me from the back of the car I was following. I felt like Peter Fonda in *Easy Rider*, and I was well and truly in my Rambling Man element. I knew I was heading into tornado country, but so far the weather was fine and I was cruising comfortably towards my first pit stop: the Launching Pad restaurant, for a wee bit of sustenance. Towering above me on the roadside was a 28-foot-high helmeted plaster man in a green spacesuit, holding a rocket ship. 'Gemini the Giant' was one of the giant 'Muffler Men', huge fibreglass figures that became tourist attractions across the USA. Gemini was purchased in 1965 for $3,000. Must have been worth it. He was visible for miles, so he was bound to draw travellers to the diner. There are quite a few of these giants along Route 66. Once the Interstate Highway was built, people had to do everything they could to attract people back to Route 66. I suppose that kind of adver-tising happens all over the world. I once saw a big sign that read *'Stratford – home of New Zealand's only glockenspiel'.* Now, that surprised me. I had thought every town would have one.

I love all those giant roadside objects you see in America.

There's a giant lobster near where I live in the Florida Keys.
And in LA, where I used to live, I always enjoyed catching sight
of the giant doughnut, especially when it had been raining,
because it looked as though it had been dunked. Spotting other
giants became my road trip game. I saw one casting a shadow
over the Palm Grill Café in Atlanta, Illinois, wearing blue pants
and a red shirt and holding a giant hot dog. Who on earth makes
these things? Someone with an enormous studio, a massive
ladder and a head for heights. The café's owner, Bill Thomas,
gave me a slice of one of his prize-winning pies. I had the peach.
Delicious. I wish I could remember where I saw the giant banjo;
I'd like to go back for another wee look.

Full of burgers and chips, I waved goodbye to Gemini and
rode on to Godley, a former coal-mining town on the Will and
Grundy County lines. A railway line used to cross the border of
the two counties and in the 1930s an ingenious set-up was
devised by the proprietors of a local brothel – they installed it
inside a railway carriage! When police were arriving to conduct
a raid, everyone would leap out of bed and push the carriage
across to the next county, where the arriving officials had no
jurisdiction. The thought of that scene really makes me laugh.
Sex workers and bare-arsed patrons all pushing the 'house of ill
repute' away to avoid charges – I love it. The madam shouting:
'Put your back into it, Tiffany!'

———

In Springfield, the state capitol of Illinois, I visited a house that
had belonged to President Abraham Lincoln. I found out that
Abe was spoilt for choice when it came to taking a piss. In the
outhouse of his home, there were three wooden toilet pots – a

small, a medium and a large. I burst out laughing – I just imagined him sitting there on one of those pots with his big hat on. I went into his bedroom too, where the wallpaper his wife had chosen was 'busy' beyond belief – it was a brown and beige leafy design over tree-trunk-like motifs with touches of blue on a cream background. It was awful. If Pamela installed that at our place, I'd get seasick in my bed. But wandering around Lincoln's house gave me a nice feeling, touching things he touched, seeing where he sat to write. People who lived in the area in his time reported seeing him pushing his kids around in a cart, which was considered a very unmanly thing to do in those days. I like that he did that. I remember when I used to push my own children around in a double stroller and some people thought I was a bit odd too. But Rambling Men care not a jot about what others think of us. We say: 'Fuck the begrudgers!'

———

St Louis is the biggest city along Route 66, and it's where I climbed inside the famous Gateway Arch. Nine hundred tons of stainless steel and hollow inside, it's a tribute to the great hunters and explorers of the West. I rode to the top of its cantilevered tramway and stood on a platform to look out over the Mississippi River, full of all those riverboats I'd previously only seen in movies. Unfortunately, not one person asked me to spell M-I double-S-I double-S-I double-P-I – that was a massive disappointment. All those years my sister Florence and various teachers tried to hammer it into me when I was a boy, and no pay-off.

St Louis is not only the Gateway to the West, it's also the home of the most exceptional creature comfort: bidets for joggers.

'Whaaaa? Methinks you're taking the piss, Bill!' I hear you say. Well, yes, in a way I am. There's a fountain in the city centre that features a runner sprinting over jets of water. I think real bidets for joggers just might catch on, although slowing down in the middle of your marathon for a quick swish of your bollocks might be a bit of a hindrance . . . You'd at least have to have a team ready to pass you the wire brush and Dettol.

Before I left St Louis, I watched part of a historical re-enactment of the American Civil War, to mark the 150th anniversary. Scores of people had been rehearsing for months to act out the Battle of Blackwell, dressed in period costumes and armed with rifles and cannons. They made a pretty good job of it. The Confederates won that afternoon, but they seemed to be a bit light on people willing to play dead for three hours – except for one guy with a bad hangover. He willingly just 'died' face up and snoozed until it was time to go home.

———

Remember I mentioned that people who had businesses along Route 66 had to do their best to attract tourists? In Stanton, I visited the Meramec Caverns, a place they advertised as the outlaw Jesse James's hideout – although there are some questions about the accuracy of that. It's a proper cave, with several levels and beautiful formations, and it has served many purposes over the years. A Native American legend of the Osage tribe suggests the tribe took refuge there from extreme weather. When it was found to hold deposits of saltpetre – a key ingredient of gunpowder – it was mined for over a hundred years. In the 1890s, people from Stanton even held summer dance parties in the cave because it had a nice, cool 'ballroom'. Great idea! And plenty of dark corners for a wee kissipoo if the mood was right.

———

In the town of Fanning, I came across another giant on Route 66. This time, it was an enormous rocking chair. I climbed up and sat on it. I was dying to rock back and forth, but the guy who'd made it informed me that although the rockers are thirty-two feet long, they don't move. Pish bah poo . . . I was inconsolable.

———

Oklahoma, Texas, New Mexico and Arizona are the real cowboy states. They had the kinds of Route 66 landscapes I'd imagined before I came – long, straight roads as far as the eye could see and enormous skies. It's perfect for biking. I passed the Midpoint Café and continued out west, where I found some really isolated towns, many of which had become cut off when the Interstate Highway was built. 'We were raped,' said Angel, a man who became known as 'The Godfather of Route 66'. In his eighty-four years he'd seen a lot of changes. He remembered the 'dustbowl days' when the route was a dirt road, he'd seen local boys going off to World War Two . . . and he was angry. 'When the Interstate Highway opened in 1978,' he said, 'we ceased to exist. The state just bypassed us. There weren't even any signs any more.' He'd spent many years campaigning for Route 66 to be given historic status.

I can't even imagine what it must have been like for local people when the Interstate Highway opened. In some cases it was very sudden, with little warning. People who relied on passing traffic for the success of their businesses must have been devastated. When the new highway bypassed the town of Glenrio in 1973, everybody left except for one woman I met called Roxanne, whose dad had owned a gas station. She stayed on with her dogs simply because it was her home. Roxanne became the mayor, the sheriff – everything. After the mass exodus, there were no stores or services nearby. She had to travel forty miles for a pint of milk.

———

In Albuquerque, New Mexico, I was driving 300 miles from Payson, Arizona, and thought I was on the home stretch. The weather was great, the roads were smooth and straight. Maybe I became overconfident, because next thing I knew I was recovering from a busted knee and broken rib. My throttle had jammed on cruise control, and I lost control of my bike. I spun out, somersaulted and broke a rib when I hit the ground. Luckily, some brilliant paramedics arrived smartish. I was so impressed with them. Three of the five were motorcyclists themselves; they very carefully cut up the seams to get my leather jacket off so it could be repaired later. It wasn't long before I was full of painkillers and well on the mend.

That was far from the first time I'd come off. I've had tons of injuries on the road over the years, but a Rambling Man accepts that, however he travels, accidents or unpleasant events are inevitable. When he's on foot the weather could make him very uncomfortable, he could trip or fall, be attacked by an animal or a human ... The same mindset applies to motorbikes: if you're on one there will be an emergency, you'll have to slam on your brakes, and you'll end up on your arse. It's part of the gig. There's an acceptance that it isn't a car. It doesn't have a radio. It doesn't have a heater. It's uncomfortable in the rain. And that's the way it is. Get used to it. It's good.

I live in Florida now, where you're not allowed to teach kids all kinds of things in the schools – even Mark Twain's banned. Thank God the Brooklyn Library is providing digital access to all the banned books for free. They give out digital library cards so, wherever you are in America, you can have access. But although Floridians can't legally read Mark Twain, they are allowed to ride a motorcycle without a helmet or shoes. I think that's a horrifying thing. Your helmet is crucial. I know riding without one might be exhilarating, like standing at the bow of a ship with the wind rushing past your ears, but it's deadly dangerous. You should also wear goggles to stop things flying into your eyes, and something over your mouth so you don't swallow hordes of insects . . . unless you're hungry for some wee, crunchy snacks. Many types of travel are dangerous, but the point is that they get you to where you want to go. You take basic precautions then get on with it. You don't examine it too closely because overthinking certain things – that's what you gave up to be a Rambling Man.

———

Monument Valley is not on Route 66 but, once I was back on the road after my injury, I couldn't resist taking a detour 150 miles north to see it. I was not disappointed. It was magnificent. It's a place that's sacred to the Navaho tribe, and very special to me. After all the cowboy movies I watched as a boy in Glasgow, I felt like I'd already been there. In my imagination, I'd hidden from baddies above the passes, swooped down on enemies from the stony heights, and sat at my campfire gazing at the fantastic rock formations under a scarlet sunset.

Navaho tribespeople have lived around Monument Valley for thousands of years. My rib and knee were not yet healed so, when I was offered a session with a Navaho medicine man, I willingly submitted to his magic. He used sacred stones, amulets, feathers, crystals and a fire of cedar chips, and it was all deeply spiritual until his mobile phone rang. 'Hello? Yes . . . No, I can't talk now, I'm in the middle of a ritual healing . . .' After a brief conversation with a telemarketer, drummers began to play, and I was told to waft my hands over the fire. At the end of the ceremony, I was a bit taken aback by the shaman's confidence: 'Bill – you're all healed up – ready to roll again!' I wasn't sure that was true, but I enjoyed the experience, I love all that stuff. Next day my rib felt exactly the same, but I was swept away by the intensity, the mystery and the sincerity of the ceremony.

In Flagstaff, Arizona, I learned that astronomer Percival Lowell had set up an observatory and discovered the planet Pluto. Lucky bugger. If they knew the real truth about the universe, they'd put a giant teacup on the road outside the Lowell Observatory, to illustrate my own 'Teacup Theory of Space'. See, I believe we are just part of a big cup of tea, sitting on the arm of a giant easy chair in space. You can talk about myriad universes and black holes as much as you like, but I think my simple explanation is one that astronomers will come to accept in future times. You may laugh. You may say: 'Bill, when did you get a degree in astrophysics?' Well, some may beg to differ, but I believe it was included in my honorary doctorates from several of the UK's most respected universities. Yes, mark my words – I know things. I have seen the future.

———

There is nothing like the Grand Canyon. I travelled there on an old-fashioned train from Flagstaff station. Who could refuse a side trip to one of the Seven Wonders of the World in a choo-choo that runs on vegetable oil? I'd never seen anything like it. It was breathtaking. But at the same time, I didn't know what to make of it because it's er . . . well, it's the world's deepest hole in the ground. In Peach Springs, not too far down the road, I stayed in the world's deepest hotel room. I took an elevator 220 feet down to my room inside a wide cavern. It had everything I needed for a good night's sleep – a bed, couch, telly, shower, toilet, dripping limestone, stalactites, stalagmites, tacky light show . . . I wondered if there were any bats lurking around, though, so I kept my genolicas well covered till morning.

As I rode through the mountain pass out of Arizona, I started to feel the warmth of California. It felt very familiar, as I lived in the Hollywood Hills for twenty years. When I reached San Bernadino, I visited an old cinema where Hollywood film-makers in the 1930s would try out their movies. They screened *The Wizard of Oz* there before it was released to the wider public, and the local audience loved it. But MGM executives complained: 'That terrible song about the rainbow slows it down!' What did they know? Louis B. Mayer refused to cut it, and it ended up winning an Oscar.

When I got into the city of Los Angeles, a guy came rushing towards me head on, then cut across the road and nearly killed me. Crazy driver. Welcome to LA. Then the traffic slowed to a painful crawl. Los Angelinos love their cars and can't seem to do without them, so there are twice as many cars as people. The Santa Monica freeway is one of the most congested in the world – a slow-moving river of road rage. Traffic reporter Chuck Street calls it 'the world's largest open-air asylum'. He would fly his helicopter and hover above it most days, advising radio listeners where the biggest snarls were. 'I got my job by flying my 'copter up to the window of DJ Rick Dees' studio with a topless woman in the passenger seat, and he went nuts. Hired me three days later.'

I finished the tour down on Santa Monica Boulevard. It was lovely to see the ocean, knowing that I'd crossed the whole of America. Accomplishing something like that is brilliant. Whenever I finish a journey, I feel lovely. I lie in my bed with my head full of all the crazy and wonderful experiences I've had.

———

One of my favourite experiences during that Route 66 trip was spending time with the Amish people, just south of Springfield. I'd known very little about them – a persecuted religious sect forced to flee nineteenth-century Europe and self-exiled from the world since then – but they were lovely. They turned out to be totally different from what I'd expected. I had thought they'd be stuck-up and prudish with their religion, 'holier-than-thou' and self-righteous, but they were the opposite. They wore their religion like comfortable clothes – very easily. I stayed in an Amish motel nearby, but we ate in people's houses. Their food was great – meat and potatoes and a dessert of cake and custard. It was like school dinner – my favourite kind of food. I had electricity in the motel, but they still used gas lamps. They make the most sensational furniture. And they can take a joke. One guy I met, Mervin, took me out in his horse and cart. As we were leaving the house, I saw his hat on a chair beside the door. I said: 'Oh, you gotta wear the hat or I'm not coming!' He just said 'Sure!' and put it on. He told me about some of the rules in the Amish community, such as not allowing pockets on shirts, and we were asked not to point the camera much at his face, as they believed it might encourage vanity.

Mervin allowed me to drive his buggy and, as we trotted along through fields covered in wildflowers, I asked 'Where did you meet your wife?' He smiled and said, 'Oh, I was where I probably shouldn't have been.' He then went on to tell me about an accident he'd had on his farm years before. While moving a bale of hay, he had reversed over his fourteen-month-old daughter and she'd died. It was profoundly moving to hear him talk about it.

A Rambling Man's life is enriched by learning about other people's lives. Being around Amish people taught me a great lesson. On the way out of Pennsylvania we stopped off at a big McDonalds, and there was an Amish family standing about outside, looking kind of awkward. People were sniggering at them. Normally, I would have just walked away, but this time I went over to the Amish family and said: 'We just spent some time with some of your brothers and sisters, and we had a lovely time . . .' Amish people are just like everybody else; they want to be allowed to live life the way they think it should be lived.

2

I'D RATHER BUILD A BOAT
THAN SAIL IN ONE

———

—

CALL ME OLD-FASHIONED, but aren't all boats just prisons, with the option of drowning? I don't like sea travel much, although I sometimes think I would like to sail from London to New York on one of the *Queens*. Those big ships make me feel proud of Glasgow's shipyards. When Pamela and I went to visit the *Queen Mary* in Los Angeles I just wandered around marvelling at the riveting and beautiful woodwork all done by men of the Clyde. Real artists. There were so many brilliant men in those shipyards, who had incredible traditional skills. The stuff they did with their hands was remarkable – especially the woodworkers, joiners and carpenters. Their carving and fretwork, the precision of the joints, and the perfectly finished claw and ball feet on the furniture were breathtaking.

I was lucky that I joined the shipyards in the late fifties, early sixties when they were still flourishing. So many great tradesmen were still working there and we looked up to them. I remember on one ship there was an opening from the engine to the propeller shaft and it was welded inside and outside. I was just an apprentice, helping out nearby, but a master welder came up to me and said, 'C'mere. Look at that!' He pointed out all the finished welding inside and outside the doorway, which was a miracle. He said, 'That's how good you have to be.' The older men would often take the piss out of us apprentices, but they also inspired us. Encouraged us to be good at our work and take pride in it too.

That all changed when I finished my apprenticeship and became a fully fledged welder. At that point the older men would always tell you that you were rubbish, especially if you landed an envied task, like working on the hull. You'd be busy doing overhead welding on the keel of the ship, balanced on a plank, and somebody would walk past and go, 'How you doing?' You'd say, 'Okay, how are you?' They'd look at your work and they'd say, 'That's fucking crap. You'll have to do better than that.' 'Shut up.' They were always giving you a bad time, but they knew you must be good because you were working on the hull.

Thinking back about the men I worked with in the Glasgow shipyards, I realise that so many of them were Rambling Men at heart, but they were trapped in a cycle of needing money to support not only their families, but their drinking habits. I would have been exactly the same if I'd not had my banjo, the dream of becoming a folk singer, and the courage to get out when I did. It was a guy in the shipyards, Willie McInnes – known as 'Bugsy' because of his protruding two front teeth – who gave me the push I needed to get out. Willie took me to task. One day he said to me, 'What are you gonna do with yourself? What are you gonna do with all this banjo playing?' He was a guitarist himself. I said, 'I'm gonna quit and join the band and travel.' He said, 'When are you doing this?' I said, 'When the holidays start.' The summer break was two or three months away. He said, 'No you're not.' I said, 'What d'you mean, no I'm not?' He said, 'You're putting it off. It's just a dream. If you really felt like that you would do it now.' Well, I didn't do it right then, but what he said haunted me. He carried on: 'There's nothing worse than seeing someone who knows they could have got away and didn't do it. It's like they're in jail.' When the summer break arrived and I got my holiday pay, I did leave and never went back. It was a leap of faith to follow my dream of becoming a folk singer and it was also the moment I surrendered fully to the life of a true Rambling Man.

———

I grew up in Partick on one side of the Clyde and my school was on the other side in Govan so I had to cross on the ferry. I loved those ferries. They were little boats run by men from the Highlands, who would always say 'Mind yer foots now, mind yer foots!' as you were stepping off. Sailing across the Clyde, I could hear all the welders and riveters and caulkers working in the shipyards. You could even see some of them from the ferry. And all the great liners would be sailing in or out – the three-funnelled ships of the Scottish Anchor Line fleet of merchant ships and many other great ships would be coming and going. They'd be registered in Baltimore, Tierra del Fuego or Rio de Janeiro, flying all the different flags, and their funnels were all different colours. I was always thinking, 'One day, I'm going to go to the Middle East' or 'One day I'll go to Shanghai and Hong Kong.' I was desperate to escape to sea. That was the early Rambling Man spirit in me.

Those ferry rides to school were proof and a constant reminder that there was a whole wide world out there ready to be explored, or get drunk in. I've seen a fair bit of that big world now. As a matter of fact, I boast to my children that I've bestrode the world like a giant colossus. I would have seen even more if my plan to run away to sea had succeeded. There was a place in Glasgow called The Pool where you could go to join the merchant navy. On two occasions my father caught me going there and dragged me back to finish my apprenticeship.

I enjoyed other ferries around Glasgow and the Scottish Isles. The PS *Waverley* is a famous ferry, with a black and white hull and a red, white and black funnel. It is a classic seagoing passenger-carrying paddle steamer – and the last of its type in the world. Everybody loves it. It reminds them of the Clyde of their childhood. But my favourite paddle steamer was the *Jeanie Deans*. She was built on the Clyde at Govan in 1931 and I used to sail on her to Rothesay when I went on holiday with my family. The journey took about an hour, and we'd stop at Dunoon on the way, where some people would get off. The Meikles who lived through the wall in our tenement building used to go to Dunoon – in those days people tended to return to the same holiday place every year. Once the ferry was under way, my father would take me down to see the engine. It was in a massive, noisy room full of painted piping, gleaming metal, and all kinds of things whirring and clunking away. It was very jolly. There was a little café on board that sold meat pies and a cup of tea, and there would be people on deck singing nostalgic songs and playing an accordion and a fiddle.

After becoming fond of the ferries around Glasgow, I started to enjoy ferries everywhere I went. I sailed on several in New Zealand. With my trike I boarded the Arahura ferry ship that sails from Picton on the South Island to Wellington on the North Island. It's my favourite ferry ride in New Zealand because the view reminds me of the Clyde. And from Auckland, I took a ferry ride to Waiheke Island, just forty minutes away in the Hauraki Gulf. They have wineries, native forests and waterfalls there, and a fabulous sculpture walk that features life-size corrugated iron cows. I contacted the artist and bought an iron cow for my home in Scotland. I set it out to 'graze' in a field behind my house, which made my Highland cattle quite bewildered.

———

A Rambling Man is drawn to anything that gets you from A to B, and ferries are often used because they're so handy. When you walk far enough, you'll inevitably come to a water's edge, so you'll have to take the ferry across to discover new land, another adventure. It's an easy way to connect one long walk to another. But apart from convenience, ferries can offer spectacular views, and they can also be very entertaining. On those traditional, big green and yellow ferries that plough their way across Sydney Harbour from Circular Quay in Australia, you pass the pearly sails of the Sydney Opera House, go under the Sydney Harbour Bridge and glide past the colourful, smiling jaws of Luna Park, onwards to the shoreline towns. It's delightful. And, during one of my British tours, I sailed up the River Thames to my gig at the New Globe Theatre, past the gigantic sightseers' Ferris wheel – otherwise known as the London Eye. Even though the docklands have been renewed and now have trendy restaurants and office blocks, the same old smells are still there. You recognise them instantly – rope, wood and whisky. Well, maybe the whisky is more something you smell in the Glasgow docks, but it was all very familiar to me.

I thought the New Globe was wonderful. The place is identical to Shakespeare's open-air theatre and my concert there was a joy to do. It was a few days after 9/11, so everyone was a bit jumpy. An aeroplane flew overhead while I was onstage, and the crowd's mood instantly changed. I pulled my T-shirt up over my head like a balaclava and said sternly: 'Don't move! People will come among you gathering your expensive jewellery . . .' People were screaming with laughter – it seemed like a huge release, as though that took the curse off all that built-up fear. After my concert I talked with some actors who had been in the audience, and they were saying how lucky I was that I could spontaneously say such things. When they were onstage and something unusual happened in the theatre, they were always forced to stick to the script no matter how much they were dying to say something else.

The medieval layout of the New Globe is perfect for a comedian. I felt that everybody in the audience was close to me. Although, people in the narrow upper galleries weren't too comfortable because their knees were constricted. I guess people were shorter in Elizabethan times . . . I had some fun with them about that. But I enjoyed having people standing in the pit right by the stage. In Shakespeare's time they were known as 'stinkards' because by all accounts they farted a lot and generally smelled foul.

———

I think the best parts of a lot of cities are often the waterside areas. I enjoy being by the harbour in Auckland, New Zealand. Pamela was born near there, and she has scores of relatives who have Māori heritage. Her Uncle Bill showed me around once. He gave me a guided tour of the dock and I saw all the big yachts. New Zealanders are immensely fond of yachting and seafaring, and they're very good at it. On Saturdays they go out and sail. It was lovely to see all the white sails out there on the harbour. They also race each other. I've never been to a place where that was a normal Saturday activity. No drinking, no football, no pub fights, just jolly good fun on the ocean waves.

They have a very different kind of boat race near Alice Springs in Australia: the Henley-on-Todd Regatta. Australians are wonderfully irreverent, so it's no wonder they dreamed up something to take the piss out of the famous posh regatta at Henley-on-Thames in the UK. But there's one main difference: there's no water in the Todd River, so they just cut the arse out of their boats and put up a sail. Then, eight to ten people get inside each boat, hold the sides up and run along the sandy dry riverbed. It's supposed to be an annual event but, on rare occasions when it's rained, the fun is ruined and they have to cancel the boat race because the river starts to fill up. That kind of nonsense is delightful to me.

———

I like the peacefulness of travelling by barge on the British canals. I did a trip starting near Sheffield during one of my British tours and it was lovely. There's something brilliant about canals. They're non-tidal, they don't whip up in storms. Mind you, I felt a bit less than manly going up the canal on a barge called *Milly Molly Mandy*. I could have used one called *Discovery* or *Pathfinder* – something a bit more rawhide and moccasins. *Milly Molly Mandy*! God, I should have been wearing pyjamas with feet.

Although I've had some lovely trips on the water, mainly for the purpose of catching things that were once perfectly happy swimming in the deep, if push came to shove I'd rather build a boat than sail in one. That might not make sense to everybody, but it's another element of being a Rambling Man, because a Rambling Man loves creating things. Whether it's carrying a banjo around and playing music, taking apart your motorcycle and putting it back together again, learning how to build something from nothing, or painting a beautiful picture, the process of creating something is what he loves best. He can take or leave the finished product because he's perfectly comfortable with impermanence.

And anyway, like I said before, once you're on a boat you're trapped there until you reach your destination, which can be very difficult if you're a restless person. But if you're in a boat and you're fishing, then that's okay. I love fishing. I'll fish anywhere, on land or on the water. In New South Wales, I caught an Australian salmon. My fishing guide and I were the only people on the beach. It was paradise. There were some kids out surfing in the ocean at the time. Nobody knew about surfing when I was a kid. 'Oohhh look! There's a man in the sea on an ironing board.' In any case, if you'd tried to surf or windsurf in Scotland, you'd have got fucking hypothermia. But did you know that windsurfing in Scotland goes back to Victorian times? We invented it, and I can prove it: if you watch the movie *Mrs Brown*, with its very authentic depictions of Victorian times, you'll see a scene where I've just been swimming and I come out of the sea. I'm drying myself off and talking to my brother and, if you look carefully, in the background behind his head you'll see a windsurfer.

Some Australians catch worms using fish as bait. Now, that's one for the trivia quizzes: 'When would fish be used for catching worms?' Answer: in parts of Australia, they mash up dead fish, put them in the leg of a woman's nylon stocking, then dangle it over incoming waves to lure the worms, which are six to twelve inches long. Once one of them pops its head up above the water, you use a credit card to trap its head against your fingernail to pull it out of the water. I loved fishing in Darwin, in northern Australia. One part of the sky was blue with white fluffy clouds, but in another part a huge storm loomed on the horizon. It was a dramatic and beautiful thing – nature showing you how insignificant you are, which is very, very good for you. Between the sun and the storm, I caught a little barramundi. It was a perfect frying-pan size. Barramundi is probably my favourite fish in the world. It's every bit as good-looking as salmon but it tastes better. It tastes even better than haddock, and that takes a bit of saying. While I was fishing for the barramundi, a red dragonfly landed beside me. I didn't move a muscle. I didn't flinch, I didn't cry or run for cover. I didn't even wet my trousers. Because that's what I'm made of.

My dad used to say that thunder was God moving his furniture around. You get the best thunderstorms in Australia. It's wonderful to lie in bed looking out at the theatre in the sky. Yippee! At sunset in the Australian outback the sky is strewn with incredible colours. Sometimes it looks like a bad painting – one of those velvet jobs, birds sitting on a burnt tree stump, all in silhouette. Extraordinary shades of gold and black. The place looks like a fridge magnet. It's glorious. Sometimes you feel glad you were born.

———

I like fishing the Green River in Utah, USA too, and lots of other places, but the best fishing I've ever experienced has been in New Zealand. Rainbow trout are native to North America, while brown trout are native to Britain, and the New Zealanders imported both varieties. Fortunately, they thrived without becoming a nuisance. In Christchurch they have trout swimming through the centre of town. That city has changed radically in recent years. The earthquakes sorted that out. I've never been in a big earthquake, but once in Los Angeles I was bathing one of my wee girls and there was a massive aftershock. A wave shot right along the bath – *shoom!* Fuck! Amy, who was six at the time, said, 'Four point five.' And she was right!

Christchurch experienced a major earthquake in 2011. The town was beautiful to begin with, but after 'the big one' everything was rebuilt at a lower height. Instead of multi-storied malls there were lots of little wooden shops with sheltered pathways outside them – a whole new feel to the town that I like a lot. I met a Scotsman there – he was a dentist whose passion was fishing. He said, 'You like to fish, don't you?' And I said, 'Yeah'. He said 'You gotta come here and fish. You get a loch each.' He wasn't kidding. See, if you turn up at a loch, and there's somebody already casting there, you go: 'Okay, I'll go to the next one.' Fishing is a natural activity for a Rambling Man. It's free food, and you can go off anywhere in the world and set up your rod, and it's just you, the water and whatever is swimming around in it. It's okay to have a pal with you, as long as it's someone who understands the rules: talk little, and don't disturb the fish.

———

Years before I moved to the Florida Keys, I used to nip there to catch tarpon. I love fishing in the back country off Key West and Sugarloaf Key. Some people there catch fish while standing up on their paddleboard. Not this soldier. And there's a fishing lodge in Tulum, Mexico that I loved. It's a simple wooden structure on the beach that has about six lodges and they serve great Mexican food. I went there with my son Jamie. We'd get up in the morning and go out on the boat with a guide, then come back at night for dinner. There was room on the boat to stand up and fly-fish in the shallows of the sea at about six feet deep – the bonefish, tarpon and permit like to swim there. I usually release fish after catching them, but one of my fellow guests in Mexico took umbrage at that. He said, 'I've never understood the idea of catching fish with all that special equipment and the clothing and the special flies – then letting it go.' The guy who owned the fishing lodge intervened. He asked the guest: 'Do you take part in any sport yourself?' 'Yeah,' he said, 'Tennis.' 'When the game's finished,' asked the owner, 'do you eat the ball?'

———

I've had lots of favourite fishing spots in Scotland, of course; catching a wee trouty or salmon in the Highlands is such a joy. Once I went to buy some flies, and an elderly Highlander was sitting there making some beauties. He gave me a wonderful fly that he'd just made. He said, 'I'll give you a wee bit of advice. When you're tying this on your line, get out of the river. Turn your back and get right behind a tree – otherwise, the salmon will be grabbing it out of your hand!'

I used to go out fishing for the day with my pal Jimmy Kent. We'd look for the perfect place and you usually know which rivers to go to depending on what they look like – salmon rivers, for example, are dark and swirly, while trout rivers look a bit browner. There's nothing I love more than the weight of the water pressing on your legs and the wee birds singing away . . . it's brilliant. You're breathing the nicest air up there – straight off the water right up your nose. When you set out to catch a salmon in Scotland, the weather conditions will always be perfect: howling gale, blazing sunshine, snow, occasional rain, soggy underfoot or very dry – it's going to be a wonderful day! You've got to be dressed for *any* weather. I was once fishing in a Scottish stream and came out of the water freezing. As I was bagging my gear, I spotted a wee Glasgow man walking past. I was wearing a Highland bonnet called a Glengarry – just to be windswept and interesting – and he was wearing brand-new plus fours that were a weird blue that doesn't exist in nature, with socks to match. He was a mess, but he had the gall to say to me: 'If I had a hat like that, I wouldn't fucking wear it.'

———

Tasmania has fantastic trout, and the people there are smashing. Country people, but very modern in their attitude. Not fuddy-duddy. I made it on to the front cover of the fishing magazine in Tasmania because I caught the same fish twice on two consecutive days. I almost landed it and it got away, but then I caught it the following day – it still had the fly in its mouth from the day before. So, it made me quite famous among Tasmanian anglers. It's a relief to fish in the warmer seas and rivers of Australia, New Zealand, Mexico and Florida. There's only so much freezing water my bollocks can take.

———

I have an outside shower at home – it's supposed to be very healthy. There's a bit of a jungle growing around three sides of it, so I feel all Amazonian while I'm scrubbing myself. It's brilliant. I do check under the leaves first, though, to see if there's anything creeping around. The only outdoors bathing experience I've had that comes close to that was in Rotorua – an interesting geothermal town in New Zealand where two tectonic plates meet. I lay in a natural thermal spring like it was a bath, and it was very pleasant. At the time, I was wearing a hairy possum swimming costume. I saw it in a shop and couldn't resist it. Once purchased, the only place to wear it was in the water, because you can't ponce around on the grass with a furry costume on. Lying there in that warm, sulphur bath, it looked as if my pubic hair was well out of control. But it was a very pleasant aquatic experience. And so was a birthday gift Pamela once bought me – one hour in an isolation tank. It was dark inside, like jumping into a tin of boot polish. I could've had a quick fifty off the wrist in there, but I thought someone might be watching on video.

———

I wanted to catch fish when I was in the Arctic but there's bugger all available there. There is a most unfortunately named creature there though – it's called an 'ugly fish'. Aptly named. It's huge with a wee tail and a big fucking ugly head like a giant wart. The only edible bit is just before the tail . . . which is its arse. Imagine going through life being called 'ugly' and your best feature's your arse!

———

The best sea journey I ever did was one that has challenged sailors and navigators for centuries. I was making a TV series called *Journey to the Edge of the World*, where I travelled from the Atlantic Ocean to the Pacific Ocean the hard way: 10,000 miles over the top of the world through the fabled 'Northwest Passage'. It was a fantastic adventure, and I'm well pleased I got to do it. For centuries it was ice-bound all year round and nobody could get through but, since the ice has been melting in recent years, there are a few weeks in the summer when you can travel through. It was very treacherous; a journey straight out of the *Call of the Wild*-type novels that inspired my boyhood adventure fantasies. I went to some bleak places.

I started off in Iqaluit, which is in the Canadian territory of Nunavut. I had arrived just in time for lunch, which was Muktuk – beluga soup – with a main course of caribou burger. That was my first clue I wasn't in Kansas any more. Suitably fortified, I flew over the icebergs to Baffin Island, home to a community of Inuit people. The place is covered in snow most of the year. You look around and everything is white, but when the snow melts a bit you can see all the things just lying around outside the stores and people's houses and you think to yourself: 'The white was nice.' They have a sign just outside the township that says: 'Road to Nowhere'. No surprise – the area is popular with courting couples. It reminded me of when my kids would come home late at night and when I asked, 'Where were you?', they would say, 'Uh . . . Nowhere.' I've noticed that in many American cities they have bars called The Office. Same idea.

Next, I flew 180 miles north to Pangnirtung, an old whaling town on the shores of the Cumberland Sound. It was a centre for the Hudson's Bay Company, which traded with local Inuit. Whalers exported whale oil, and it was a huge industry until the whale population declined. I was so surprised to meet an Inuit man who played Scottish music on an accordion. He obtained the instrument in 1973 from Hudson's Bay Company. When the place was flourishing, Scottish whalers spent the winters here, and they brought their music and their instruments along with them. There's a Scot or two in every corner of the world. We're a nation of Rambling Men.

As I began to trek to the Iuituk National Park I reached the beginning of the Arctic Circle. It was a patchwork of snowfields, tundra plains, ice flows, and mountains towering above me – all perfectly reflected in shining, clear lakes. I was told this was polar bear country and was hoping they'd hand me a rifle so I could protect myself. Unfortunately, they just taught me how to use bear spray. Armed with my trusty aerosol, I strode out into the wilderness . . . and fell on my arse crossing a freezing stream. It was peculiarly quiet. A feeling of big emptiness. As I climbed into my tent that night, I recited my usual bedtime ditty to myself: *Goodnight my dear and sweet repose. Lie on your back and you won't squash your nose.* I just added my own ending: *And keep your bear spray handy.*

———

The next day we soared over amazing fiords in a helicopter, and the Turner Glacier, which is 2,500 square miles of ice. It is currently melting due to global warming, and it makes a jolly noise while it's at it. Later that day I met an Inuit elder named Abraham with a laugh that sounded terribly like that melting glacier. I was beginning to feel like I was on another planet – a stunningly beautiful one. Deep inside the Arctic Circle, on Igloolik Island, people live in a traditional way. You know when people in the UK say, 'The wind was Arctic . . .' Well, they have no idea. I had to be fitted for a made-to-measure sealskin suit because my London-bought winter jacket was considered well under par for the Arctic temperatures. The woman who made it measured my feet using the kitchen floor tiles. She chatted away while she measured my entire body, saying a lot of stuff I didn't understand. But, loosely translating, I think she said: 'You have wonderful manly arms that could kill a caribou with one squeeze.' Well, at least I could probably frighten an animal to death by jumping out from behind a tree dressed like that.

Suitably attired, I sailed round a few icebergs to a historic ceremonial site at Igloolik Point. In 1822, while on an expedition to find the Northwest Passage, Captain William Parry and his men were iced in here. All hell broke loose when one of the crew started seeing a local married woman. The headman of the town asked the local shaman to cast a spell to get rid of the sailors and, the story goes, the ice broke the very next day. Parry's men set sail, but a 36-year-old man named Alexander Elder had not made it through that winter of 1823. When they tried to dig his grave, they broke ten pickaxes. No wonder the Inuit didn't bother burying people; they just set them on an iceberg.

Pond Inlet on the north coast is the stopping off point for the Northwest Passage. The icy wilderness there is beyond belief. It's the most unforgiving place I've ever seen. I went hunting in the area, but it was a long way from tiptoeing over the moors to bag a few grouse. I felt guilty being present during the seal hunt, even though I had all kinds of rationalisations: seals are not an endangered species, there are 5 million in Arctic waters, and the Inuit eat them to survive. As I was watching from the boat for a seal to stick his head up, I had to remind myself it was part of the culture. I was all for them doing what they've done for centuries and knew that they need to be allowed to kill their quota of animals, but I was still a bleeding-heart hippy liberal, and it was tough to accept. I learned a lot about myself in the Arctic, and about how the real world works. Some people shoot things, while others go to the butchers.

Hundreds of people died attempting to find the Northwest Passage. I was travelling in the footsteps of pioneers and dreamers, of great Rambling Men. I was meeting people who lived in the wilderness on the very edge of the world. That journey was made mainly by ship and, in all, I sailed 10,000 miles through the Arctic Ocean to Vancouver Island. It was supposed to be summer, but it was bloody freezing at the weather station in Resolute Bay. There was just tundra – nothing much grows there. Flying in there is very dangerous. I saw a wreck of a Canadian Lancaster bomber – my favourite aeroplane of all time. This one had crashed in 1950 during a weather report run. Miraculously, only one person was injured and nobody died. By contrast, the first people who arrived there came by boat 100 years ago. The Tuli people, who were ancestors of the Inuit, travelled by sea across Alaska to Greenland. I saw the remains of their boat, with a wee lemming living inside under the whale-bone arches.

———

For the next part of my journey, I joined a Russian cruise ship with a hundred passengers aboard to take me north. It was only the second cruise ship to visit that area in six years – and my first cruise ever. First up, we did a drill for 'the unlikely event the ship started sinking'. We all had to go to muster stations and practise for the end of our lives. I asked: 'Will we all fit in that wee lifeboat?' Someone answered, 'Well, it will be nice and cosy.' I said: 'Well, I'm just looking for who I'm most likely to eat if it all goes pear-shaped.' Nobody responded. I'd obviously created an uncomfortable atmosphere. Over the loudspeaker, the captain continued ominously: 'Next time this happens it will be for real.' I went to the bridge to talk to the captain. 'This ship is a Class B ice ship,' he explained. 'We can't break the ice, but we can manoeuvre around it.' Good to know.

In the mid-nineteenth century, an explorer called Sir John Franklin attempted to find the Northwest Passage, but it was a complete failure. The British knew it wasn't realistic, but they sent him anyway, with 129 men in two ships – none of whom were ever seen again. In the winter of 1845 they got stuck on Beechey Island, where many of them died. I was supposed to go ashore at that very island, but three polar bears were spotted just before I arrived, and I was not about to step ashore with some hairy oaf waiting to scoff me. From the ship I could see the place was grim. Awful and cold, even though it was summer. One of Franklin's men, John Torrington, was buried there. In 1984 a post-mortem was carried out and they found that he'd died of lead poisoning from eating the canned food. Those men were like astronauts. They were very well equipped. They had a library aboard their ship containing 3,000 books. The ship also had a carpentry shop, a blacksmith and a shooting range. Even with all that, I couldn't imagine what it was like being stuck in a place so desolate, wondering if you'd ever get back home again.

The colours I saw while cruising the Arctic Sea – the intense blues of the water and the sky, the shining silvery-white of the ice – had a Disneyland theme park feel about them. It was wonderful to look at. But I'd think twice about coming here without sheepskin underwear, and I'd definitely think twice if

I didn't know whether I could go home again. As we sailed further north, we found that ice was blocking our way. The wind had blown massive pieces into our path, so we had to change our route. I love that even now, it isn't a given that you can get through the Northwest Passage. The captain tried to get an icebreaker to come but it wasn't available, so we set a course through Peel Sound. We had the luxury of an ice chart, but I was trying to imagine what it would have been like without one, in a wooden square-rigged ship, like Franklin and his men, with their tinned food killing them. We sailed to King William Island, where Franklin's men set up camp after they had to abandon their ships. One hundred and twenty-six wretched men died, one by one, in this frozen wasteland. The end of the Franklin Expedition was at Victory Point – the most miserable place on Earth. Ten years later, search parties shocked Victorian Britain by reporting evidence of cannibalism. It must have been hellish. This is not a sport for the white linen trousers and a picnic hamper, I'll tell you that.

If you go to Westminster Abbey, you'll read that Franklin discovered the Northwest Passage. He did not. Fifty-three years after Franklin's disappearance, a Norwegian guy called Roald Amundsen was first to get through it, on a three-year journey from 1903 to 1906. He did it the right way. He had a smaller ship with a shallow draft, and only six crew members. He stopped off on King William Island, learned the language, got to know the locals, dressed in furs and did his homework on the place. There are still people there with Norwegian blood, so the Norwegians obviously had a few cuddles with the local women. Nevertheless, Amundsen said the best thing you can do is leave the Inuit alone to live the way they always did. Armed with local knowledge, he managed to navigate the whole way through to

Vancouver Island; but he couldn't have done that without a discovery made by a Scottish explorer called John Rae. Travelling by dogsled, Rae explored a southern route without which no one would ever have got through. I think we should send a stonemason down to Westminster Abbey to set the record straight.

———

On our trusty cruise ship, we finally made it through the Northwest Passage. After disembarking, I crossed the Arctic Ocean by plane to an isolated community that sits on the very edge of the continent – Tuktoyaktuk. That's not a word to be attempted with loose dentures. But what the people who live there call themselves is even harder to pronounce – TUKTUUYAQTUUMUKKABSI. Do they think they're Welsh or something? For the final phase of my journey, I hitched a ride on an ice truck along the Dempster Highway – an ice road – with a guy who regularly made a 5,000-mile round trip delivering fruit and vegetables. Many years ago, I picked up a hitchhiker in Scotland who had the temerity to tell me he didn't like my comedy, so I stopped the car and let him out. There, on the blizzard-prone edge of the Arctic Circle, I was extremely careful with what I said to Dan the driver. Later, I hiked in the Tombstone Mountains with a gun-toting woman called Lolita. We climbed to an unnamed lake. I suggested they call it 'Lake Thank God – I Couldn't Have Walked Another Step'. Think it will fit on the road signs?

Eventually, I arrived in Dawson City, which is in Yukon, Canada. In 1896, where the Yukon River joins the Klondike River – a favourite fishing location for indigenous people, who'd lived there for thousands of years – George Cormack found gold. It became the hottest place for gold in the world. It was known as the Klondike gold rush. Some took out 50 pounds of gold a day with dredgers. Gold mining is still alive there. I had a go at panning. It's hairy-chested men's work, and I can't say I enjoyed it – there's something about wet underwear that never appealed to me. I'm a big Jessie. My father used to call me a 'Jessie' all the time. In Scotland that means you're not manly and you can't play football very well. He would say: 'You're a big fucking Jessie!' Years later I was in Hawaii and there was a restaurant called Jessie's. Pamela took my picture in the car park next to a sign that said *'Parking for Jessie's Only'* and sent it to my father.

Standing in the middle of a breathtakingly beautiful countryside, I staked my own gold mining claim by sinking a post in the bush and calling it the 'Billy Claim'. I wish I'd given that a bit more thought. I missed the opportunity of a lifetime to name it something more fuck-youish, like: 'Don't Even Think About Panning Here Or BIG FURRY THINGS WILL BITE OFF YOUR BALLS!' By law, no one can move that claim for the next hundred years. I suppose I should go back and try a bit harder to find a wee nugget. But it's not a region for Jessies. I had the worst evening at the campfire – the guys talked about bears all night. I was trying to sleep but I was convinced there was one nearby that would think I was a tasty dinner, so I didn't get a wink.

———

The next day, I met the most impressive woman I've ever met in my life. A true Rambling Man named Nancy. She ran a fabulous, remote ranch guesthouse near Telegraph Creek. The town of Telegraph Creek used to be like Las Vegas, providing supplies and entertainment for people when they came through during three gold rushes, but now it's a ghost town. Fifty miles further down a dirt road, Nancy – who was five foot two inches tall and seventy-five years old – managed a 480-acre ranch all on her own. She was so confident and capable I was sure she trapped grizzly bears and wrestled them to the ground. She took me out on her boat with an outboard motor, then drove me in her tractor to the guesthouse. Her driveway was four miles long. My arse was making buttons. I was absolutely flabbergasted by her strength. I watched her split logs for winter with only one blow of her axe to each one. She regularly gutted moose, and she used her rifle to scare off wolves and bears. 'Once I saw a bear coming down from the hills into my yard,' she told me. 'I knew the kids were playing outside, so I ran and got my rifle and shot it just as it was crossing the creek there.' Nancy was completely self-sufficient and I loved her contentment and enthusiasm. She took me for a walk around her land. It was a wonderful place with breathtaking scenery. Then, that night, she made me moose stew with cabbage and beans for dinner and a cakey treat afterwards. I wanted to ask her to marry me. When I went up to bed, I found a note she had left for me in my room. At first, I thought: 'Hello! She feels the same way!' But it wasn't like that. The note said: *You must have a good sleep. If you are cold, it is miserable. I advise warm nightwear and warm socks should be worn to bed.* My kind of hotel. I asked her, 'Don't you think this way of life could get too hard for you?' She just scoffed. 'I'm not a quitter.'

It was almost time to go home, but before my trip came to an end I had to travel 600 miles south to the Native American settlement of Gitlax̱t'aamiks for a traditional sweat lodge ritual. 'What does that involve?' I asked. 'Oh, you're going to be boiled and steamed.' They weren't kidding. Inside the sweat lodge was a steam bath, created by the intermittent pouring of water onto hot rocks. It was pitch black, apart from the glow of the red-hot lava rocks that turned white as they burned.

At times, it was so hot in the sweat lodge I felt like I was sitting on the inside of a volcano just as it's about to erupt. There were five sessions, each one hotter and more intense than the last. At one point, a girl brought in frozen fruit in her Thermos. It was delicious – eating the frozen fruit in the room of fire.

People were saying prayers, chanting and talking openly about all kinds of physical and emotional experiences they'd had, while the temperature continued to climb. People believe the ceremony cleanses you inside and out. It removes negative energy and creates better balance in your life. Another bucket of water would be poured on the rocks and a wave of hot steam would suddenly roll through the circle. The chanting grew louder. Everyone was sweating but, even though it pushed me to the limits of my endurance, it felt peaceful. Some of the men talked about being taken away by the authorities when they were children, happy children, living in their culture. But the authorities insisted they had to be taken away to strange schools and houses all over Canada, where their language and culture was all beaten out of them. Their parents had had no say in it. They weren't allowed to speak their language or eat their food. These grown men were crying at the fire about how much they missed their language and the company of their peers. They said it had all been stolen and it wasn't going to come back. A big muscular man was crying like a baby because of what had been done to him. I was so moved by it.

I shared some things too. I admitted that I sometimes thought I wasn't grateful enough, that I didn't take time often enough to appreciate the life that I'd built, the people around me, and the world in general. It's difficult to explain exactly why or how, but the sweat lodge had a profound effect on me.

The following day I chopped a tree down in a place called Horsefly. A man named Leonard Cecil and his crew of loggers dressed me up in orange clothes and a helmet, and declared I looked the part. We were to fell trees infested with the destructive pine beetle. Dangerous business. Over the course of my trip to the Arctic and through the Northwest Passage, I had started to become such a ruffty-tuffty. I had been a nice tree-hugging hippy when I left home. I thought my wife would barely recognise me in my new Grizzly Adams persona. That's what the raw outdoors will do to you. 'Chainsaw cuts tend not to hurt,' said Leonard. 'I've had shaving cuts that hurt more.' His face was covered with scars, but I didn't like to ask which activity he was worse at – cutting down trees or shaving. He showed me how to use a chainsaw, and I felled a big tree. Naturally, I hugged it first. The sheer power of it falling was extraordinary. Quite a rush. I love the camaraderie of working men, but I was disappointed that nobody shouted 'Timberrrrr!!!' And not one of them was wearing a tartan shirt – I was inconsolable.

———

I ended that long journey at Friendly Cove in the Pacific Ocean. I had travelled 10,000 miles in ten weeks. Friendly Cove was discovered by Captain Cook when he was looking for the Northwest Passage in 1776. At that point, after his exploits in the Pacific Ocean, Cook had retired, but a substantial prize of £20,000 had been offered to the first group through, so the British Admiralty decided Cook was their man. But, like many men before and after him, Cook found zilch. The centre of a doughnut.

For me, it was an incredible journey. I'd had a unique chance to see some extraordinary places in the Arctic summer. There were some lovely people there. I'd never felt helpfulness or welcomeness like that anywhere before. They were people who knew who and what they were and wouldn't live anywhere else. And I'll never forget that constant terror of polar bears, or the powerful feeling of being absolutely surrounded by all that water and ice. It was profoundly humbling.

———

Humans don't really understand the power of water, do we? I pulled a hairdryer out of its bag in an English hotel, and there was a wee tag on it. I thought it would say *'Return to front desk. Don't even THINK about putting this in your suitcase, you big fucking long-haired oaf! And if I had hair like yours, I would carry my own fucking equipment.'* But you know what it said? *'Don't use this in the shower'*. I thought: 'Who is this notice for?'

3

STRUCK BY LIGHTNING

—

—

I'M NOT VERY vocal in my sex life, I'm just quietly grateful. But there are other situations where I have been known to let out a scream – like on an aeroplane, when it takes a downward turn halfway across the Atlantic Ocean and feels like it's plummeting seaward . . . 'AHHHH!' But that kind of soiled-trousers panicking is all behind me now. I stopped being afraid of flying once I realised that if the plane crashes from its cruising altitude you won't feel a thing. Once you get above maiming height you can relax.

It's normal to be wary of being in the sky. All Earth-dwelling creatures feel the same. I saw two caterpillars watching a butterfly. One said to the other: 'You'll never get me up in one of those things.' Au contraire, I love being up high. It's just the plummeting to your death I don't fancy. I have never suffered from vertigo – probably because I got used to working on suspended planks in the shipyards. And I survived a massive fall from the deck of a ship when I was a teenage apprentice welder. 'Lucky Bill' they called me after that. In the seventies, I suffered from reverse vertigo: when you're *not* high you're frightened.

New Zealanders seem to specialise in jumping from great heights. In Auckland, I sky-jumped off the tallest building in the southern hemisphere. I felt like an idiot, though, because despite having been in the parachute regiment I landed like a bundle of washing. Kersplat! Fubble fubble! 'Owww! Fer Pete's sake . . .!!!' Times may change, but standards should remain. In Queenstown, New Zealand I discovered one of the best laxatives known to man – bungee-jumping from a suspended platform into a gorge. And I did it naked. I walked into a sturdy building standing above a gorge. The people who operated the jump met me at the entrance, then showed me where I could leave my clothes. I then had to walk naked into a bigger room where there were about a dozen other people, some of whom worked there, but others who were just their relatives who'd shown up because they were nosy. A lovely wee blond boy with Down's syndrome was there. He had a beautiful face. He said: 'Hello!' I said 'Hello!' He said: 'Where's your clothes?' I said: 'Over there. I like to jump without my clothes.' He said: 'Why?' I said: 'I dunno,' and he laughed. I went through the outer door, then along a passageway, to where there was a narrow platform like a diving board. They put the gear on me there. It was a harness similar to parachute gear, with webbed straps and buckles, and two straps that came up the side of your thighs and round your waist. I was instructed to walk further forward to where the platform widened a bit. I stood on this twelve-inch square platform, and I was thinking: 'I hope I don't fall!', which didn't make any sense because I was about to fall anyway. Then they told me to just wait a bit until I was ready, then tip myself forward. There was no pushing and no machinery. It was just silent. Quite holy. Just you and the world. You free-fall for ages, feeling the wind rushing past your face. You can see all the mountains and the trees as you tumble down. I remember seeing a group of large boulders coming towards me – because I was leading with my face – and thinking: 'This wasn't such a great plan. Not the best idea you've ever had, Billy.' But then I began to like it. Eventually I could feel a sort of 'give' as the line had reached its length, and then it wheeked me back up in the air. 'Wheeehhh!!!' I bounced two or three times. That was a lovely feeling. Freedom. And then the line went slack. After that you hold on to the rope while they bring you up slowly and you can just step onto the platform again when you're level with it.

But when I got back onto the platform I felt as if I'd told a dirty joke to nuns because I was naked and they weren't. The filming was over, so the charm and bravery went out of it, and I felt I was just a naked guy showing off. All the people in the room wanted me to sign stuff – photos, books, albums – but I said, 'You'll have to excuse me – I feel awkward. I want to get my clothes back on.' And I could see some of them were checking out my willy, so I just stared back. 'Yes? Can I help you?'

———

Compared to bungee-jumping, flying in an aeroplane is a doddle, although I'm very afraid of the moving floor at the airport. I choose to walk alongside it. I've tried for many years to get on to one of them without looking like a real prick. Most people manage it perfectly. They just walk on, talking to someone. 'Aye, Bobby, aye . . .' and on they go. Maybe I think too much about it. I try to adjust my speed as I approach it, which usually ends up with me tripping. Lurching. It's a real dignity-stripper. Sometimes, when you lurch, your elbows come up. You didn't ask them to do that, it just kind of happens. And sometimes you even make a wee noise. A wee, astonished noise: 'Oouughhhh!' I'm sure there are guys sitting in a room somewhere controlling those things. They're watching their video monitor and they see you coming along. 'Oh

– there's one!' 'That was a cracker, Willie – she grabbed the wean, hahah!' 'He smashed his carry-out there!' Bastards. 'Make the banister go a bit faster, Bobby . . .' 'Ya bastard you!' The only way to get rid of the embarrassment is to pretend you *meant* to lurch. But that means doing the whole journey like that.

Another flying embarrassment is sleeping on an aeroplane with your mouth wide open, and sometimes your tongue falls out. Although, drooling down your jacket is a very attractive look, thinks not you? And men tend to hold their willies when they sleep. Now, there's a very good reason for men holding on to their willies while they sleep. It goes back to the beginning of time, when cavemen had to protect themselves and future generations from testicle-eating wolves that roamed the land. But female passengers nearby say: 'Look at that! Typical! Can't keep his hands off it!' Because women are consumed with penis envy. If women had penises they'd never leave home. We should swap parts for a wee while. The streets would be deserted.

Flying is not the natural state of a Rambling Man, but it's another means of getting somewhere. If you want to travel to India or play a gig in Canada you'll have to fly. Aeroplanes are unlikely to be a Rambling Man's preferred form of air transport; I think his ideal mode might be an airship. I'm intrigued by them, although I never saw one; they had ballrooms and everything. I used to have a great postcard that Ian MacKintosh the banjo player sent to me years ago. It showed people in a street in Germany pointing in the air at a zeppelin. Ian had written his own caption on a balloon coming out of a guy's mouth: 'Taxi!' Nevertheless, commercial aeroplanes certainly offer the Rambling Man opportunities for new adventures. He can travel far greater distances and visit exotic parts of the world he'd previously only read about.

The first time I flew overseas was when I was a welder in the shipyards. A fellow welder had told me he was enjoying being in the Territorial Army (now known as the Army Reserve). He said he had a great time, sometimes getting a few weeks off work to go on exercises abroad. I thought, 'I could do with some of that.' The whole idea was well in sync with my Rambling Man sensibility, so I went along and joined the parachute regiment. We trained on weekends and I discovered I rather enjoyed jumping out of towers, balloons, and eventually helicopters and aeroplanes. The first seven flights I ever took I never experienced a runway landing, because I always jumped out of the aircraft with my parachute. Once I qualified, I was eligible to go on the trips abroad. It was all good fun – marching, shooting a rifle, sleeping outside, lighting fires and eating tinned food. They were like glorified versions of the camping trips I'd enjoyed as a teenager. Then the big day came when I was given three weeks off welding to go to Cyprus for more intense military training. We flew there in one of those big military planes where you sit against the walls, then parachuted out into pitch-black night and landed in the fields below. We had to find our regiment and then march through the night to a camp base where we began various games and shenanigans. It was a blast.

On the way home from Cyprus our plane got struck by lightning. I would never have known that it had happened, though, because there wasn't the loud noise, searing flash of light or terrifying ball of flames you'd expect to witness if your plane got struck by lightning. There was just a wee bang, like a door being slammed. Mind you, I wasn't next to a window, I was in the seat opposite the toilet door, so I couldn't see outside anyway. At the very moment the lightning struck the plane, I happened to be eating treacle pudding. I had become quite partial to military food – stuff you can travel with, like biscuits that become porridge after you soak them. The treacle pudding was particularly good, and it came in a tin with a kind of twisty razorblade thingy with which you opened it. I'd just got it unsealed when the plane suddenly dropped hundreds of feet. *Whooosh!* My treacle pudding flew out of the tin. I'd never seen a pudding levitate before. Quick as a flash, I shot out my arm and caught it in my fist. We lost pressure, so the oxygen masks came down from the ceiling, but, naturally, I ate my pudding first, then licked my sticky fingers. Well, you wouldn't want to get treacle pudding on your oxygen mask, would you? Then eventually the captain announced we'd been struck by lightning, and there was a big hole in the fuel tank, so we were going to have to land in Malta to fix it. I didn't care. Another day off from the shipyards suited me just dandy. And anyway, it was in Malta that I got the only shave I've had from a barber. I love being shaved with a cut-throat razor. It's one of the most luxurious things a man can experience. And at the time it was a godsend, because I had spots. For me, it was a nightmare trying to shave around my acne.

During one of my tours of New Zealand I visited a museum in Auckland and saw a replica of the plane built by Richard Pearse, the first man to fly – he did it before Americans. He actually managed to get airborne but, on his maiden flight, he crashed into a hedge on his farm. I once flew in a biplane not much bigger than his. It was during some filming for an Australian TV programme. I was in the front of this tiny two-seater plane, and the pilot, Bob, got in the back, which was a bit of a heart-stopper. I was thinking: 'Oh shit, how's he gonna drive it?' I was pretty nervous. I asked Bob: 'Do we have parachutes on board?' He said, 'No. We're gonna come back.' I hate it when people in charge of your safety think they're comedians. I tried to make friendly conversation. 'Have you ever parachuted out of a plane?' I asked. 'No,' he replied. 'I can't think of a good reason to abandon a perfectly serviceable aeroplane.' Then he asked me a lot of questions about how I became a stand-up. I wanted to say: 'Eyes on the road, Bob,' but I told him I was a funny folk singer, and eventually I graduated to doing solo concerts. 'That's the pinnacle for comedians,' I said, 'like flying solo is for you guys – only slightly more frightening.' Then, just to show me what frightening really is, he started to loop the loop. Oh fuck. My stomach was okay, but my arsehole was pulsating. Bob said, 'The first time I went solo, I chundered everywhere.' I managed to squeak: 'Me too.' After we landed, he said the camera battery had died while we were up there, and would I mind going up to loop the loop again. 'You fucking kidding me?'

———

There are times when I wish people didn't know anything about me before they meet me, but that rarely happens. They usually already have an opinion of me, but I never know exactly what that is because it can depend on whether they've seen me performing or just read scurrilous things about me in some newspaper. I don't read newspaper articles about myself, so I have no idea what strangers might believe I'm like – a funny man, a mad beast or the devil incarnate. Anyway, I once glimpsed an article that started: *'Connolly's technique is extraordinary'*. Having no real insight into what exactly my 'technique' was, I thought: 'I better read the rest. Maybe I can find out.' The article continued: 'He leaves the subject for hours on end and returns unerringly to where he left off.' That was bullshit. I never return to exactly the same place; in fact, 'quite near' does me fine – which is exactly why I never became an airline pilot. Know what I mean? There'd be a news report: 'The plane landed quite near Heathrow' . . . *Bump! Bump! Bumpity Bump!* 'This is your pilot Captain Connolly speaking. We have landed quite near the airport. Everybody off! There you go . . . What you complaining about? It's London, isn't it? Well, Londonish . . . Take your luggage and your family and fuck off. Out! How can I take off on a ploughed field with you fuckers all sitting there? On your way.'

When you fly as much as I do, you're bound to have the odd glitchy-poo. I was once going to meet Pamela in California. Prince Charles and Diana were about to get married, so we were going to watch the royal wedding on TV together. Well, it was before there was Netflix. But shortly after leaving Hong Kong something went horribly wrong with one of the engines. I don't know what the trouble was, but the plane was banking alarmingly and ejecting fuel to try to land. Most of the passengers were freaking out, but there was a guy in front of me who was very calm. He said: 'It's nothing! You're perfectly safe, I've crashed six times!' He was a helicopter pilot in Vietnam. And then the pilot announced: 'We're gonna put down in Tokyo.' When we landed, fire engines were lined up all along the runway, but nothing bad had happened and everybody was thankful to have made it. Then the pilot gave the passengers a bewildering choice. He said: 'You can choose to fly in the same plane to Los Angeles, OR we can get you a new plane if you want but it will take a long time to get another aircraft here and set it up. The old one is all ready to go . . . but the choice is yours!' I would have thought it was a no-brainer to wait for a well-serviced, mechanically sound aircraft – even if it made you late. But the pilot had extraordinary powers of persuasion, because in the end people just shrugged their shoulders and we took off again in the same plane. We didn't do it sober, though. Everybody crammed into First Class, and we had a wild party. People were crazy, drunk, out of control. I think the drinks were free. There was a guy running round trying to shag everyone. He had a sticker on his forehead that said: '*I may be old, but I still get hot!*' I don't think they kept a 'No Fly' list back then.

Another time I was flying from Canada to Los Angeles. We'd just settled down after taking off when a guy jumped out of his seat and shouted, 'We're all gonna die! We're all gonna die!' He was insisting that we land. 'We're all gonna DIE!!!!' People were shitting themselves and the staff were coming down the aisle looking for a doctor to treat him. But the only 'doctor' on board was an American actor who was the star of a medical TV series. He said: 'I'll have a go.' He was very cool. He just said, 'Sit down man,' from his seat. Everyone turned round and realised who he was. He had a deep, authoritative voice and, surprisingly, that did the trick. The panicked guy sat down, and it was all over.

Using the right words in the right way on the right occasion can make all the difference. Once a guy came up to me on a plane from Glasgow to London. He introduced himself as the son of Pastor Jack Glass, a man who plagued me for years by calling me a blasphemer, protesting outside my concerts and even throwing missiles at me. His son had obviously taken up his case. He said: 'How would you be if it was Judgment Day and God was to judge you now?' We'd just taken off. 'Fuck off!' I hinted. The guy next to me said, 'My God – is that what you have to put up with?' But by then Glass's son had dutifully fucked off. Yeah, it pays to get straight to the point.

But besides the odd uncomfortable moment, I've had plenty of wonderful trips by aircraft, like flying up to Galway in Ireland to see where my family came from. The sea has been battering that region for many centuries. I looked down and saw all the huge stones – nature's Lego – just lying all over the place. It was right near where John Alcock and the Glaswegian Arthur Whitten Brown crashed after achieving the first non-stop trans-atlantic flight. Alcock had been in the Royal Naval Air Service and dreamed up the trip when he was a prisoner of war during World War One. He and Brown took off from Newfoundland in June 1919 in a twin-engine biplane. They made it over France towards Ireland, but after ice, hail, dense fog, and a failed engine they went off course and were forced to crash-land in a bog in Connemara. Brave buggers.

————

As I said, if a Rambling Man had a favourite mode of transport in the sky, it would probably be an airship. I've always fancied taking a trip in one of those, but they stopped taking passengers after the 1937 Hindenburg glitchy-poo. I've heard they might be coming back as 'greener' alternatives to aeroplanes though. There's a spy blimp in the sky near where I live, probably keeping an eye on Cuba or watching for drug runners from South America, I don't know. It looks mysterious and kind of wobbly. I don't really like anything that wobbles. I once had an experi-ence with a wobbly UFO that gave me the fright of my life. I was on the deck of our house in L.A., playing my banjo, and out of the corner of my eye I saw something moving along the valley beneath me, weaving in and out of the buildings. I didn't know what it was. Then it turned and started to advance towards me. It turned to face me, looking like a weird airborne reptile, and I became paralysed with fear. Was it an alien spaceship? There must have been the same kind of quizzing going on in my brain as First Nations people experienced when they first saw European ships appear on their horizons. Eventually it dawned on me that it was a huge bunch of helium balloons tied together. They had been attached to an archway at the entrance of the second-hand car dealership in Ventura Boulevard, down in the valley below. They'd been put there to attract customers, but local pranksters had cut them loose and they'd floated away. A good time was had by all – except me. Big Jessie.

I like flying in helicopters. They're lovely. Much safer than their reputation. I flew over New Zealand's stunning Fiordland in a helicopter. It was real Grizzly Adams land. Beautiful. If I was being tortured with cigarette burns, I might even admit it's nicer than Loch Lomond. They have fat turkey-like birds there, and they use glove puppets for artificially rearing them without their mothers. They even showed me how they do a kind of Punch and Judy show featuring ferrets and weasels, to teach the birds about predators. Now do you believe I've seen it all?

I've also parachuted from a helicopter. You jump from a sitting position with your legs dangling outside. The guy in charge comes along and touches your shoulder to signal it's your turn to jump. Helicopter life is so different from aeroplane life. More relaxed. You'll be flying along over, say, Wellington, and the pilot will say 'Fancy a coffee?' 'Yeah.' He'll land behind a coffee shop, and you'll have your macchiato then get back in the helicopter and off you go again. If you need to pee, you come down and get behind a tree.

———

My worst experience 'needing to pee' was in Mogadishu, Somalia when I was filming Comic Relief. My first task upon landing there was to appear in front of the local bigwigs and explain what we were doing there. I had been bursting to pee in the wee plane, but they only had those little narrow-neck bottles, and I would have been pissing all over my legs. I just had to hang on till we landed, but my back teeth were floating. The second we landed I shot out of there like a rocket. The director of the Comic Relief programme said it was the funniest thing he'd ever seen – three rows of dignitaries standing on the tarmac waiting to greet us, and I came crashing right past them holding my willy. We didn't dare film it, though, and it was a good thing they all laughed.

In Mogadishu, you pass kids going to school and they say: 'Good morning!' And you say: 'Good morning!' back, and then they say: 'How are you?' and roar with laughter. They are lovely kids. If you can get the ball off them – which is hard cos they're very good – they may let you play football with them. They're exceptional at tricks, like bouncing the ball up and down and catching it on their heads. Delightful people.

———

Once when I was in Australia, I left my trike in Sydney and travelled to Coober Pedy by plane. This was a sensible choice. If you travel by road, you'll see a sign that says: *'Next Service 257 Miles'*. Fuck that. I also went by plane to Alice Springs, which is a nice-looking place. We went for dinner in a Swiss-Indian restaurant. It was great – in the morning, we didn't know whether to fart or yodel. And we flew over Ayers Rock, as it was called then, before it changed to Uluru, which seemed like the end of the world. Suddenly you can get some perspective on prehistoric things. We're just a tiny blip in time compared to this stuff. Coober Pedy lies halfway between Adelaide and Alice Springs and is the opal capital of the world. I've never been so hot in my life. The people there have dug their houses underground. Some are carved out of solid rock and the temperature inside is very cool. They even had an underground community swimming pool that was so cold they had to heat it. My hotel in Coober Pedy was underground, and it's something you have to get used to. One night, I put the light out and I couldn't find the bed. I'd never been in such Stygian darkness. It got quite scary. Took an incredibly long time to find my bed. Or even to find a wall. I just had to keep edging around until I found a wall, then I walked round the wall until I found my bed.

Coober Pedy really feels like Australia. I had an overwhelming desire to break into 'Waltzing Matilda'. The golf course green there is definitely not green. It's sandy-coloured. You carry a wee square of AstroTurf with you around the course to tee off. And you don't need to worry about ending up in the sand because you're permanently in the sand. Frankly, I've never understood golf or golfers. It seems to me when the Scots invented the game many years ago, it was a joke. 'Eh, Jimmy . . . Hit this wee ball made of feathers with this stick.' 'You're taking the piss, right?' It was never supposed to be taken seriously. And there was nothing in the rulebook about V-neck pullovers with wee lions on them and matching tartan trousers. Golfers should be fucked and burned. Give golf courses to the homeless.

———

Planes are undoubtably necessary for some trips, but if you only ever travel by plane you can be anaesthetised from truly experiencing places. I was in a bar in Invercargill, the southernmost city in New Zealand, and a Scottish woman was pouring the drinks. She said: 'This is a great place.' And I said: 'Yeah? What – New Zealand?' And she said: 'Invercargill. It's a great place. And I've been all over the world.' I said: 'Where have you been?' She listed the places one by one. 'I've been in Dubai, Singapore, Sydney, Melbourne, and I've been up there in Auckland!' And it wasn't until later that I discovered that's where the plane stopped on the way to Invercargill. She hadn't really been anywhere. But then, sometimes people say to me, 'Have you ever been to Istanbul?' and I'll immediately say, 'Yes!' but then I'll remember that I haven't seen the city. I've just spent hours during a layover traipsing around that massive airport trying to find my fucking gate. Just when I finally thought I was on the right track I saw Pamela careening recklessly towards me on one of those carry-on suitcases that turn into a scooter. She had just bought it at an airport shop because she'd hurt her knee and couldn't walk, but she didn't know how the brakes worked. She was a massive danger to everyone in her path and I had to run and catch her; that was MY Istanbul.

During the gig, I told the Invercargill audience that I'd seen an ugly mob running down the main street and other people joining them from side streets, chasing one poor guy. 'Come back!' they were shouting. 'Call yourself a Kiwi? Fucking traitor.' So, I asked them: 'What's the story here?' 'Oh, he's the guy who hasn't seen *Lord of the Rings*!' I said I was frightened to confess I hadn't seen it either. I said: 'To be perfectly frank, I like movies with people in them!' But that was before I played the King of the Dwarves, Dain II Ironfoot, in *The Hobbit: The Battle of the Five Armies*. After appearing in that movie, I loved the whole concept . . . although I still haven't read Tolkien. I'm far too busy for the likes of him. I'm fully occupied with my Proust and Dostoyevsky.

Trying out my friend Willy Kelly's Matchless motorbike while camping in Arbroath as a teenager.

London to Brighton charity bicycle ride.

A puncture-fixing crew in Bali.

These Hong Kong 'ding ding' trams used to be in Glasgow. We called them the 'caurs'.

Fab biker jacket made for me for my World Tour of New Zealand.

Lie back and think of Scotland: July '87 – five months before his marriage in Fiji – the bridegroom practises his favourite Kama Sutra position.

Arriving in Adelaide on my purple Harley-Davidson three-wheeler.

Whenever I was on the road I sent my wee girlies faxes I created.

Accident waiting
to happen.

Coober Pedy, Australia.
To stay cool, people live
inside the rock.

A human polar bear magnet enters the Arctic Circle.

There's a Scot in every corner of the earth: I listened to Simeonie Keenainak playing Scottish whaling music in the Arctic Circle.

Testing my hearing aid in a massive echo chamber overlooking Turner Glacier.

A wee stroll in the wilderness is good for the soul.
Tombstone Mountain, Yukon.

Is this what you press to make it work?
Visiting the Meramec Caverns in Stanton, Missouri.

I met the fabulous
wrestler, Cassandro,
in El Paso.

Driving the Amish buggy with Mervin in Arthur, Illinois.

The view had me on the edge of my seat. The 'World's Largest' rocking chair in Fanning, Missouri.

Fry and fly: a monster of engineering, powered by chip fat, which took me to the Grand Canyon.

Halfway along Route 66 in Adrian, Texas.

Happy in Monument Valley: a quick tune on my banjo (*left*) before taking a wee stroll among the sandstone buttes (*below*).

4

WHAT DO STETSONS AND HAEMORRHOIDS HAVE IN COMMON?

—

I LOVE TRAMS. They go at a proper speed, where you can see the world passing by gently rather than whooshing by too quickly. When I was a boy, I would ride the Glasgow 'caurs' – as they were known – with my father to the centre of town, going to the Sunday market known as the Barras. They made a lovely sound: *clankity clankity clank*, and they smelled of metal and hot rubber. The conductors were great, especially the women. They were funny and dead cheeky. They told riddles and jokes and everybody on the trams enjoyed it. But in the early sixties the Glasgow trams disappeared. People had complained they were clogging up the place and making the traffic more congested, so they replaced the trams with trolleybuses and eventually with diesel buses. The women workers kept up their patter on the buses. A guy would get on and say: 'I'd like a single to Drumchapel please.' The conductress would say: 'We're not going to Drumchapel!' 'But it says "Drumchapel" on the front!' She'd give him an old-fashioned look and say: 'Well, it says "India" on the tyres but we're not going there . . .!'

After the Glasgow trams were decommissioned, some of them were imported into foreign cities. When Pamela and I visited

Hong Kong in the 1980s, it was exciting to see the same trams I'd known in Glasgow when I was a boy in operation around the island. They were all dressed up in primary colours with advertisements for Fuji and Mitsubishi and renamed 'ding dings'.

Melbourne has the biggest urban tram network in the world. I was on a Melbourne tram once and someone came up and asked me for my autograph. Shortly after, another curious person came up and said, 'Excuse me . . . are you a film star?' I said, 'Nope, I'm not a film star . . . and I don't think I ever will be!' What did I know? I never thought I'd end up making fifty films.

I have a lot of affection for Melbourne, and I love the Australian wry sense of humour. Melbourne is considered to be the epicentre of gunzeldom. A gunzel is like a trainspotter – someone who knows an awful lot about things that nobody gives a shit about. You'll be taking a picture somewhere, and you'll hear a voice say, 'Is that a B13 you've got there?'

'What?'

'Your camera.'

'I dunno.'

'I think you'll find if you open the door and look under the hinge it will say B13.'

'Oh fuck – so it does.'

'I thought so. It's exactly the same as a B15. They changed it in 1964. The B15's a slightly longer hinge and a slightly narrower door. They found that by shortening the hinge and widening the door they had much easier access. A man called Thompson designed it. Died in 1974.'

You're starting to edge away. 'Is that right?'

'Aye. I personally have the B17 . . . a longer door and the switch inside that transfers the gunzel power to the ignition switch is slightly . . .'

They're obsessive. If you don't want to stand there for another hour, you have to be blunt and cut them off. 'Excuse me . . . Fuck off, will you?'

Trains are just as beautiful as trams. I made a TV show once called *Tracks Across America*, where I travelled through Minnesota, North Dakota and Montana. It was a brilliant journey. The only trouble is the '*whooo hooo*' noise – they do it all night long. The train driver sits there in the front doing it at every crossing. It's ridiculously frequent: '*whooo hoooooo hooo*'. After a while you get used to it, though. Your mind seems to just tune it out.

Taking a shower on the train is difficult, although I found out some people get really creative with it. The guard on my train said he stumbled upon three guys showering together. He said, 'I won't tell you what they were doing, but one of them was upside down.' Impressive athletics, thinks not you? I'm not sure what a guard was doing walking in on three adults taking a private shower, but apparently there was a complaint about the noise . . . '*Whooo Hooo*'???

There's a seat in the shower, but it's the toilet. You can sit on it while you're showering and it helps to keep you balanced. Otherwise, every time the train lurches you're jolted back and forth. In one way, that should help you soap yourself, but it's also a heart-stopper. It's actually a very well-functioning shower. Perfect temperature. You have to prise yourself out of the bathroom cubicle to get your towel, but it's adjacent to your sleeping compartment so, theoretically, nobody should see your bare arse. Everything about a moving train is potential comedy. It's lovely watching people trying to keep their balance. They have a peculiar gait. They walk up the train and, when they squeeze past you, you both giggle. Make a little remark. They get thrown sideways onto things and fall into people's laps. It's very jolly.

I was accompanied by a crew of about six people, and the whole journey took about three weeks. The beds were wide enough for two people, and they folded down off the wall – they're called Murphy beds. There was a restaurant car that operated on a rotation system. They gave you times when you could go for your dinner because they couldn't fit everyone in at once. It was good food. I used to like the bean roll – I had it every night. It sounds terrible, but it was delicious. We would travel overnight to the next state and then we'd get up and go filming.

———

My favourite state along the train route was Montana. The people there were cowboys . . . and Native Americans. And the occasional yodeller. I met a true yodeller in a town called Shelby. He stood playing his guitar for me next to the railroad tracks, and when he started singing I could barely believe my ears – it was amazing, a real joy to listen to. He called it 'fancy cowboy yodelling', something he learned from his father when he was a boy. He told me his father would yodel old cowboy songs whenever he was happy. As he started to play another tune, I heard a *'whooo hooo'* in the distance, and a train flew past us as we sang together:

> *All around the water tanks,*
> *waiting for a train*
> *A thousand miles away from home,*
> *sleeping in the rain*
> *I walked up to a brakeman,*
> *to give him a line of talk*
> *He says 'If you've got money,*
> *I'll see that you don't walk.'*

———

Many people I came across in Montana were ranchers and people who worked on farms. Some were rovers who went from farm to farm – true Rambling Men. Hard workers. It's a tough existence, and it's simple in many ways. I visited a radio station where there was a whole programme dedicated to swapping stuff. People swapped farm machinery and all sorts of things I'd never heard of – swapping a 'gizmo' for a 'gekkmo'. It was a bit of a dull programme, but it gave me ideas. If they had a swap station in Florida, it would be a great way to get rid of my floral Hawaiian shirts. I'm sick of them. And I'd swap my wife for a Winnebago. I'd love a Winnebago. I always thought it would be a great way to live, just pulling into lay-bys and sleeping there for the night, then meeting some more people and living your life in true Rambling Man fashion. Unfortunately, I don't think you can do it like that any more – people with car parks don't let you stay the night.

During that train journey I came across real hobos. 'Hobo' may not be a politically correct word any more, but I care not a jot. It's just another name for a Rambling Man . . . or maybe a subset of Rambling Men. In any case, the word is widely used. Just look at any high fashion website today and you'll notice it's in vogue to have expensive, unstructured, designer 'hobo bags'. They're based on the carryalls used by men and women who travelled around America seeking work during the Great Depression. I've been using that kind of casual shoulder bag for at least fifty years, so that must make me a truly fashion-forward Rambling Man, thinks not you?

Anyway, there are far more derogatory words for travellers than 'hobo' – like 'derelict', 'bum', 'vagrant' and 'beggar'. What's the difference between a hobo, a tramp and a bum? Some say a 'hobo' is a migratory person who *wants* to work, a tramp will work, if necessary, but a bum doesn't want to work at all. Some say the word 'hobo' is an abbreviation of 'Homeward Bound', which referred to soldiers returning home at the end of the American Civil War. Others say that the word comes from 'hoe-boys', referring to farmhands who would travel around carrying hoes and other tools. But regardless of the terminology, a common theme in the history of these particular subsections of Rambling Men is that they traversed the country on boxcars, which are train carts. So, you see, trains and Rambling Men are inextricably linked.

The filmmaker Martin Scorsese made a film called *Boxcar Bertha* about a famous hobo, loosely based on Bertha Thompson's autobiography. She was no stranger to hopping freight trains. *Boxcar Bertha* – what a great name! In the USA there are many famous hobos with wonderful names, such as: 'Frog', 'Hobo Lump', 'Connecticut Tootsie', 'Grain Car George' and 'Angie Dirty Feet'. There was also a musician called Boxcar Willie, who sang songs like 'The Lord Made a Hobo Out of Me' and 'Hobo Heaven'.

> *A hobo's dreams are just like wings of angels,*
> *They'll take him anywhere he wants to go.*

———

A guy I was at school with called Jimmy Logan became a hobo in South Africa. He was an interesting guy. People were always telling him what he *could* have been. He was funny and he probably could have been a comedian. He was a good football player and had gone to South Africa to play for the Bloemfontein Rangers football team, but he never made it back home. I met up with him because his brother asked me to try to find him when I was touring there, and I did. 'What happened, Jimmy?' I asked. He said: 'We were getting fifty pounds a week, and wine was fifty pence a bottle. I didn't want to come back. I was happy.' He just rode the trains there from town to town. When he died, he was on the front page of the Glasgow newspapers. They quoted him as saying 'I'm a hobo and Billy Connolly gave me a hundred pounds.'

During the filming of *Tracks Across America* I met another modern hobo, who was a lovely musician. He played the guitar and sang very well. He travelled around playing music and working on ranches, then sent money home to his wife. He seemed happy – that's what I liked most about him. He told me that at any given time, there were between one and two hundred hobos on the train. But he also said things had changed in the hobo world since 9/11, and that it was getting harder and harder to jump on a train and get inside a car.

And I heard there's a punk sect that travels by train from time to time. A punk hobo sect! They are horrible to come across, apparently. Violent. There have been grisly hobo murders. *Clockwork Orange* stuff. Whenever there has been a rise in unemployed people and economic hardship in certain areas of the USA, hobo numbers – and competition for a spot on a train – increase. I felt a longing to join them, because I still have a very romantic notion of train-hopping, even though it wouldn't be easy at my age and in my state of health. But the hobos weren't in First Class drinking free margaritas. They were in the Sore Arse department. Anyway, with my Parkinson's disease, I sometimes shake so much I can just sit in my living room with my eyes closed and pretend I'm rattling along on a freight train.

——

Some of my favourite memories from filming the *Across the Tracks* series are of meeting Native American people at county fairs and festivals. I loved seeing them dancing in their traditional dress, with all the gear. Wonderful. I like the chants, but I'd always wondered what the sounds meant. I asked a tribesman if it was words or sounds, and he said, 'Just sounds.' He said, 'It's like a mantra, but we know the sounds. It's like a language, and we have to preserve it.' He told me a parable about two boys who were fishing from a tree that had fallen down into the river. One was sitting at one end and one at the other, but they weren't catching anything. An elder walked by. 'Are you boys catching anything?' he asked. 'No,' they said, 'we're not getting any bites.' The elder said: 'I'm not surprised. You're not doing as you were told.' He took the line and slung it into the water, then pulled it back in a jerking, rhythmic way, while chanting: '*Heya, haya hi heya haya . . .*' He got a bite immediately. He taught the boys that, to be a successful fisherman, you're supposed to fish in the rhythm of the song.

It was fantastic to see all the tribespeople dancing, stamping their feet. '*Heyyyyey yey ye.*' Suede fringe, feathers and beads. I loved it. We had gone early to where the festival was happening, but no one was there yet. So we hung around eating hamburgers, thinking 'Oh, it's just another fair day in America.' Then suddenly we heard '*Dumdododo Dumdododo Dumdodododo.*' They danced in through the gate. '*Hey . . . eyeyeye*'. Brilliant. I was transported to the Western plains, sitting on my horse, and the 'Indians' surrounding me. When I was a boy, I used to go and see 'Cowboy and Indian' movies and the people in the cinema used to cheer the 'Indians' and boo the cowboys, because Glaswegians identify with the underdog. The cowboys represented authority. The 'Indians' had better horses too, and fabulous jewellery, especially the Navaho stuff. I met Navaho people in New Mexico, and the Hopi there as well. I've been wearing a silver and turquoise Navaho bracelet for years.

When we stopped off in El Paso, a lovely wee town along the train route where lucha libre is very popular, I met a man who wore the most fantastic outfits. He was a type of Mexican wrestler known as an *exótico*, who often fight in drag. Cassandro was wearing glamorous white and gold boots, a coiffed, blond mullet and silver eyeshadow, and a long black and gold cape that he whipped off as he approached me to reveal a white and black bodysuit covered in diamanté. It was quite a look. Cassandro explained that he'd been an *exótico* since 1987 and experienced a lot of discrimination for being an openly gay man in the sport of wrestling. He received a lot of aggression from women, who thought he was beating up their husbands or boyfriends, and was once stabbed in the stomach by a woman wielding nail clippers. He told me that his father had been absent for most of his life, but a few years ago they'd finally managed to talk and became close again, and he now considered his father his best friend. It was lovely to hear that.

There's an amazing boot store in El Paso called the Rocketbuster. They do handmade custom cowboy boots – the best in the world. The boots have been created according to themes. I would like to wear the 'Eddie Cochrane' boot on one foot and the 'Buddy Holly' boot on the other, although you'd probably have to buy both pairs. I already have some of their boots and they're my favourites. My 'Devil With The Blue Dress On' boots are my best ones. I saw them in a magazine from Canada called *Cowboys and Indians* that was all about jewellery and clothing. I got in touch with Rocketbuster over the phone and they said: 'Send a template of your feet.' I sent it immediately, and they sent me fantastic boots – black with white stitching, featuring red flames and skeletons in blue sombreros drinking tequila from bottles, and female devils in blue miniskirts with red hair and sunglasses, playing blue and red maracas. They were better than I even imagined. And they fitted brilliantly. Pamela got me some nice boots from Rocketbuster. They have my own foot tattoos replicated on them. You have to be careful not to go too far with the cowboy look, though. You know what Keith Richards said to me about the Stetson? 'What have Stetsons and haemorrhoids got in common? Every arsehole eventually gets one.'

5

FUCK OFF ALL OF YOUS!

—

I HAVE NOT had much luck with cars. My first car exploded on the very first day I drove it. It was a yellow Morris Minor with a vase of artificial flowers stuck to the windscreen. I loved the look of it, but it was a terrible mistake of a car. At the time, I didn't yet have a full licence, so my friend Billy Johnston came with me to 'supervise' while I drove to my gig. On the way to Fife I suddenly noticed there were clouds of smoke pouring out of the exhaust and bits of burning stuff flying out and landing all over the road. Might have been a bit of a hint that all was not well with it, eh? But Billy wasn't paying attention and I was a bit of a novice, so I just kept driving. Shortly afterwards, it burst into flames. 'What's the hand signal for "I think we're on fire"?' I asked Billy. 'I think it's THIS!' he replied, as he wrenched open the door and scarpered out on to a housing estate.

Yeah, my experience with cars has not been so good – although I have had some beauties, like my hot-rod – a pale-blue 1937 Ford roadster. It was a joy to drive. I miss cruising around LA in it. The hot-rod is the quintessential American car. I love the rebelliousness of the whole idea: a fast car made by young people out of older cars like Ford Model Ts, or any of the Ford family.

It was such a great idea to take old vintage motors and strip them down. People would take all the rubbish off them, like the hoods, roofs or mudguards, to leave the car open at the top and the wheels exposed. To finish it off, they usually give the car a jazzy paint job. When you do that, you turn it into the same animal as a motorbike – a serious 'fuck you' machine.

Cars are not the preferred vehicle for a Rambling Man – he'd rather walk or head out on two wheels – but hot-rods embody the spirit of a Rambling Man. There are no rules or limits to hot-rods, you can do anything you like to the car, choose any shape, colour or design. You can add flames to it or flowers. 'Hot-rodding' is not just about the car, it's more a certain kind of attitude and lifestyle. Towards the end of my 'Route 66' trip, I visited a garage in Pomona, Los Angeles, where there is a family hot-rod business. I really admired the extraordinarily flashy vroom-vrooms they had created. They even do colourful, shiny, souped-up boys' bicycles – I would have killed for one of those when I was a teenager. I would have cut a serious dash on the Great Western Road on one of them. I wouldn't have looked the part though, sitting on it in my woolly vest, gloves on a string, and balaclava to keep my head from turning to ice. Back then, every Scottish boy had a balaclava. If a Martian came down, he could be forgiven for thinking the place was populated by seals.

The city of Los Angeles gave birth to hot-rods. People would race them on the Strip on Friday nights – and they still do to this day. The California Highway Patrol has reported that South Los Angeles and Van Nuys in the 'Valley' are hotspots for illegal street racing, which increased by 27 per cent in 2021. Well, it was a perfect outdoor activity for the pandemic, thinks not you? In Pomona I saw the best car I've ever seen in my life. It was a shiny, fire-engine-red contraption with red plush inside the doors and it was named 'El Patrote' – 'The Pimp'. A guy was sitting inside it, working the hydros to make it do that jerking, coughing thing with the noise to match. Brilliant.

————

I had an iridescent green Volkswagen in the eighties when I was living in London, that came to a bad end. I flipped it on the road when I was driving to Cornwall to film a scene for the movie *Water*. It was late at night, after my fortieth birthday party. I know what you're thinking, but I had stopped drinking by then. No, I fell asleep at the wheel and shot into a pasture. Another motorist saw it and called the police, who came to rescue me. I bashed my head, lost consciousness, and I woke up at the hospital. My eyes were red instead of white – it was a unique look. I had to wear sunglasses to finish the filming. They released me to go to the set the next day, provided I return to a London hospital and have a check-up. The show must go on. My pal Dennis Dugan was appearing in the movie too. He was also an artist, and he created a whimsical piece for me – a framed 'L' plate with scratches and bits chipped off it. I had lost my 'Advanced Driver' status.

The place where I crashed was near Weston-super-Mare – John Cleese's birthplace. I was glad I hadn't died there; I'd probably have had to share the headlines. *'John Cleese's Hometown Site of Billy Connolly Fatal Accident.'* Bollocks to that. It never ceases to amaze me that some Americans mistake me for John Cleese. It happened to me several times when I was alone, but it also happened to me when I was in a Manhattan restaurant with Eric Idle. A woman came up to the table and addressed me:

'It's wonderful to see you! You're a hero of mine.'

I said, 'Thank you very much.'

Then she said: 'Where are the rest of the guys?'

I said: 'Which guys?'

She said: 'The rest of the Monty Python gang.'

I said, 'You think I'm John Cleese, don't you? Well, I'm not.'
Eric was pissing himself.

She said: 'No, no, you're not going to get away that easily.'

'I'm not John Cleese. I'm Billy Connolly.'

She refused to believe it. 'Oh, I get it. You're incognito . . .'

When we were leaving, she waved and called out cheerily,
'Bye, John!'

———

Since my concert tours were always stressful, I had to choose
the most relaxing ways to travel. At first, I had drivers or roadies
to drive me around, but some years ago I took to driving myself.
I took great joy in that, especially in a great, reliable car. I could
stop and have a scone, drive to the gig on the B roads, and listen
to music I liked without having to consider anyone else. I'd play
some folk music, a tape one of my pals had given me, some Bob
Dylan . . . or listen to Radio 4. *Desert Island Discs* was always
great value, because there was such a mad variety of musical
styles in each programme – say, opera, hymns, pop and an
obscure Hungarian csárdás – all lumped together by Joanna
Lumley or some other well-known person.

Of course, cars do break down, and that can be a monumental
problem if you're on your way to a gig alone. But then, I've also
had a few mechanical breakdowns on my trike over my years
of touring. In Australia once my chain came loose and wrestled
the mudguard into a torn mess. I had to have it all taken off,

straightened out and put back again. I was on a very tight touring schedule, but the mechanic in Melbourne was extremely accommodating. He said: 'I'll do it in an hour.' And it was a brilliant job. Lasted the whole tour. There was no need to get a new part – it was perfect. Australians are very good like that.

For the Rambling Man, a car doesn't have the advantages of a bicycle or motorbike; you can't smell the fields or feel the air rushing round you – but you can relax, chat away on your 'hands-free', and take a couple of pals with you. I did a film with Murray Grigor called *Clydescope* where I cycled the length of the River Clyde. The actor John Alderton followed a similar route alongside me in a Range Rover. The idea was to show two different ways of touring Scotland – the posh way and the cheap way. We started at the source of the River Clyde down at the borders in Lanarkshire and went all the way up to Rothesay. It was great. I composed songs about the Clyde and sang them with my banjo:

> *Oh, when you're lonely and tired inside*
> *Grab a steamer, sail down the Clyde*
> *Go to Arran, Millport or Rothesay Bay*
> *No kidding, it's a magic way to spend a day.*

——

The next time I toured Scotland on film, it was in the spring of 1994 and I was the one driving a car. It was for my *World Tour of Scotland* series, and it was a brilliant trip. I'd do my concert at night, then get up in the morning and film a piece about the area I was in. Easy-peasy. In those days I had a lot of energy. I'd think nothing about filming all day then performing at night.

That was the first time I'd toured Scotland for many years. I played forty concerts in a range of concert halls, from eight nights in the 2,000-seater Usher Hall in Edinburgh to a night or two at the tiny hall in Orkney that could barely accommodate 200 people. I drove a maroon Range Rover for most of the tour, but the first leg was aboard Caledonian MacBrayne ferries – they're a large ferry service that cruises around the islands and peninsulas on the west coast of Scotland. We sailed from Gourock down past Rothesay, where I used to go on holiday when I was a boy, with Largs on the left-hand side. For many people, this is the best part of Scotland – the Scottish isles. Great Cumbrae, the Isle of Islay and the Isle of Arran ... picturesque places that are still remarkably unspoiled. I love going to Arran. It has beautiful, jaggy hills where you can do some serious climbing, or more gentle hikes if you prefer. Or you can just ramble around. There are little cafés in towns like Blackwaterfoot, Brodick and Whiting Bay – it's a smashing place. Tourists visit Arran because it's considered to be a kind of Scotland in miniature, but I go there because my pals live there. Playing a small venue on a small island where your pals live is a great way to get your heart started. I have to own up, get out my gear and have a little play with them. In Arran, I'm not a big touring artist; I'm just a Rambling Man with my banjo and my autoharp, playing along with my pals in a wee pub. I love that.

While I was aboard the ferry, I remember pointing out a particular ship that was sailing down the Clyde. It was a lovely sight, with its wee yellow funnel that stood out against the background of the beautiful, misty hills of the Scottish isles. Most people would never have known it was the sewerage ship that left from Shieldhall in Glasgow. It sailed to a granite islet in the mouth of the Firth of Clyde called Ailsa Craig – also known as Paddy's Milestone, because it's halfway between Belfast and Glasgow – and back again. Originally the ship just transported Glasgow's sewerage, but then the people who ran it had a brilliant idea. 'We're already sailing in the most picturesque part of Scotland – Rabbie Burns was born around there for heaven's sake – why don't we take day trippers along for the ride?' So, people started going aboard for sightseeing excursions from Glasgow – West Kilbride, Gourock and Largs – trying to ignore the fact that beneath their feet there was a hold full of Glaswegian excrement.

———

The fault line that separates the Highlands from the Lowlands starts in Loch Lomond in the town of Stirling. The places to the north of Stirling are a wee bit less accessible than the other holiday places. They attract a different kind of person – people who want to be active on their holiday. There are a few crazy people who will ski off-piste in all kinds of treacherous conditions and, even in summer, tragedies befall those who make the fatal mistake of being unprepared for a Scottish mountain. But I just love the Highlands. I had a house there for a few years, near Strathdon in Aberdeenshire. My family wasn't crazy about the winter conditions there – it got dark around 3 p.m., and the snow got so deep there were high red marks on poles to show the snowploughs where the road was supposed to be – but a couple of the summer months were unbeatable. Whenever we could, we'd invite pals to stay, visit the Highland Games, and dance a wee jig. We had some magical times, roaming around the countryside laced with wee sparkly streams, having a laugh with wonderful local people, visiting ancient, ruined castles and cycling along high, treeless ridges in such stark landscapes you felt you were on the moon . . . I miss it.

Glasgow was next on my filming list so, once I arrived back on the mainland, I drove to my hometown. Glaswegians have a mixture of Irish and Scottish backgrounds, which has made them funny. Many people I knew when I was growing up just killed me with their patter. You'd hear such funny lines, especially in pubs. I once heard a guy say, 'I wish I was born rich instead of handsome.' The older men in the shipyards were the best. They lightened their load by being hilariously dirty, and ruthlessly teasing the apprentices. They'd send us off on wild goose chases. 'Hey, Billy! Go to the supply store and grab me a bubble for my spirit level will ye? There's a good lad . . .' and they'd ask us for some 'tartan paint', 'sky hooks' or 'a long stand'. We were too green to know they were just taking the piss. They'd call you a 'stumor', which is a Glaswegian word for a silly person, a person of not much intellectual worth. I don't recall hearing that anywhere else in the world . . . except for in America, where it's the name of a politician. Maybe he spells it Stumore or Stu Moore, but either way he'd last about a fortnight if he came to Glasgow.

I took the camera crew to Dover Street, where I was born on our kitchen floor. My poor mother was in agony giving birth to me – I was 11 pounds and 4 ounces. School cap, badge and blazer, and thumped my head on the floor too, which might explain everything. Our flat was three floors up in the tenement building but it doesn't exist any more. About twenty-eight years after I was born on that kitchen floor, I was in the house of a guy called Andy Moyes and I had a very bad hangover, so I was at his kitchen sink, splashing water on my face. By chance I looked up and, through the window, I saw the house I was born in suddenly falling down. Fuck! I was a hippy then: '*Aghhh*! This *means* something!!' And being a big-headed wretch, I said: 'Where are they gonna put the plaque now the building's away?'

We drove to the oldest house in Glasgow – Provand's Lordship. It was built in 1471 and is still standing solid as a rock. I fully expect that at Hogmanay 2030, it will still be there looking exactly the same – maybe with a few fairy lights, and with a bit of sing-song floating out of it.

Hogmanay used to mean a great deal to me. Once upon a time the real tradition of Hogmanay – taking a lump of coal and a dram of whisky to neighbours' houses and receiving a dram in return – was a common tradition. Glaswegian working-class houses were mainly tenements, which made it very easy. If you went to the top flat and had a drink or two, then came down the stairs and had another drink at each flat, you'd be pissed by the time you got to your own landing. Nowadays you knock on the door and say, 'Happy New Year!' but in the old days you'd bring a lump of coal and say: 'Lang mey yer lum reek' (May you always have a fire in your hearth). You'd put the coal on the fire, then they'd give you a drink and you'd toast each other. People would also say: 'When the mouse crawls into yer oatmeal, may he not come out empty-handed.' I especially enjoyed Hogmanay because it isn't a religious time, it's a celebration – a big party where you bond with your neighbours . . . or, as we say in Glasgow, get pished with them.

I remember one Hogmanay when I was around seventeen and lived in Drumchapel, I was at a house party full of people I had just met that night, and ended up passing out on someone's bed beside all the coats. When I woke up, I only had one shoe and had to make my way home like that, reeking of booze and tobacco. Back then, Hogmanay was a small family or community affair, but nowadays they have massive indoor ceilidhs in some cities like Edinburgh attended by thousands of people, and the lovely old-fashioned thing of going to your neighbours is almost gone. It used to be normal to walk along the street, hear a party up in a building somewhere and go up and just join in. I used to do that a lot with my friends. I used to talk about it onstage too. I would mimic a drunk man trying to walk straight, with the takeaway booze in hand, and hearing the party songs. When I was younger, the parties were more like the traditional ceilidh where they would take it in turns to sing an Irish or Scottish song at the party – one person at a time. It was a hit parade of sentimental songs that all seemed to be about how much they missed Scotland and their own people. It seemed daft, because they were right there.

> *Though I'm far across the sea*
> *My heart will ever be*
> *Back hame in dear old Scotland*
> *With my ain folk*

———

When I'm away from Glasgow, I miss it. I once wrote a song about the town called 'I Was Born in Glasgow':

> *Oh I wish I was in Glasgow*
> *With some good old pals of mine,*
> *Some good old rough companions*
> *And some good old smooth red wine.*
> *We could talk about the old days*

Fuck Off All Of Yous!

And the shipyards' sad decline
And drink to the boys on the road.

Oh I was born in Glasgow
Near the centre of the town,
I would take you there and show you
But they've pulled the building down.
And when I think about it
It always makes me frown,
They bulldozed it all
To make a road.

———

Once Glasgow lost the docks and the shipyards, the heavy industries that kept the shipyards rolling along went too. They had created a lot of smoke and dust, which made the cityscape of Glasgow appear dark and sooty. When I was a boy, it looked like it had been drawn in ink. I liked it though, especially when it was wet, and the cobbles were all shiny, and there were tram-cars. It was magical. It's all bright and cleaned up now and the light has come back, so it looks more like a watercolour. People seem to love it more – but I'm a bit nostalgic for the old, grimy place. In the 1940s, me and Florence would play out in the streets. You could play in the streets then – there were no cars. Germans were bombing the place, but nobody gave a shit. They'd just look skyward: 'Fuck off, you German bastards!' That's Glasgow.

The reason Glasgow existed is because of the Clyde. People settled by the river, as they do all over the world. The whole of Glasgow life was there. It thrived on all the ancillary industries

– makers of rope, leather goods, tables, chairs, all sorts of things that are needed for the ships. Only two major shipyards are left now, both of which operate for the design and construction of naval ships, and there is one other smaller dock that builds car ferries. As I stood on the banks of the Clyde looking across to Govan, I remembered that right there was an old graveyard where moneylenders would sit on Sundays. You couldn't legally buy alcohol on Sundays, but there were many people who needed it every day, so scalpers sold booze at inflated prices at one table and the people who could lend you the money for it sat at another. Quite a racket. Glasgow was a very different place back then.

———

Partick and Govan residents weren't only on opposite sides of the Clyde; they always seemed to be on opposing ends of an argument and were frequently fighting. A Govan man got into a stramash with a Partick resident one night at the F & F Dancehall in Partick. He landed one on the Partick guy, then ran off, with the Partick man and all his mates in hot pursuit. He ran to the dock where the Govan ferry crossed the Clyde and was relieved to find the ferry was a distance of about five feet out from the dock. He took a running leap and just managed to scrape himself on board. He climbed up on deck, turned round to face the angry Partick mob and yelled: 'Fuck off all of yous!!' But the captain touched his arm. 'I'd take it easy if I was you – we're coming into dock.'

> *And that great old place I miss so much*
> *Has seen much better days,*
> *And still we talk about it*

Fuck Off All Of Yous!

As we go our separate ways.
Oh, but Glasgow gave me more
than it ever took away
And prepared me for life on the road.

———

Before I left the city, I headed to Glasgow Cross and stood beneath the tall structure where they used to hang people – the Tollbooth Steeple. The last recorded hanging here was a famous thief called Billy who wore a red bonnet. A massive crowd turned up to watch him hang and, when he dropped into the pit, his brothers ran up and grabbed his legs and swung on them so he would die quickly. Just before he was hanged, he threw his red bonnet into the crowd and there was a big fight for it. The guy was like a rock star.

I love those expansive gestures that were once popular among rock performers. I never threw anything into my audience, but I knew guitarists who threw their picks and drummers who tossed their drumsticks at the end of the show. Nowadays it's the audience that throws stuff onstage . . . teddy bears, bras, lollipops, sex toys . . . it's madness. Call me old-fashioned, but the last thing I'd want flung at me while I'm onstage is a pair of glow-in-the-dark handcuffs.

———

The Scottish knight Sir William Wallace was hanged, drawn and quartered in Smithfield in London in 1305. He was accused of high treason but protested: 'How can I be guilty of treason when England is foreign to me?' Some people have that tattooed on their arms. At Stirling Castle there's a lovely monument to

William Wallace. He fought for Scottish independence and defeated the English at the Battle of Stirling Bridge. They have his sword in there – a double-handed claymore. It's some size. Did you see the movie *Braveheart*? What the fuck did Mel Gibson think he was doing? Or the writer of that movie who thought it was okay to present something so unbelievably historically inaccurate? Oh well, I know I'm not too good at dates or accuracy either . . . I might as well tell you that ancient Scottish kings were crowned by sitting on the Stone of Destiny naked, with a rasher of bacon round their necks. Only the first part of that is true. The Stone of Destiny was originally there in Perthshire, Scotland, but it was taken to Westminster Abbey, then stolen by 'Scottish Patriots' in the 1950s. Its replica – or possibly the real thing (nudge nudge wink wink) – sits at the entrance to Scone Palace. Ancient kings of Scotland have been crowned by sitting on it ever since King Kenneth MacAlpin – whose daughter was Queen Margaret. There were no carriages, mounted horses or hordes of flag-wavers; it was a simple ceremony – you just came up, sat on the stone, and – lo and behold! – you were the king of Scotland. The crowned kings would always have earth from their own land in their boots so they could technically be standing on their own soil. They'd empty it out afterwards – which is why there's a mound nearby.

Stirling Castle, where that Wallace monument is, is set in very fertile country that goes all the way to Loch Lomond. The smell of tea and wood burning, and the ever-changing light dancing across Loch Lomond, will hook you for the rest of your life. No wonder bus parties come up along the one-lane road. Clean air, mountains, lochs and forests – all for free. And there's nobody there. It's the most extraordinary country.

Fuck Off All Of Yous!

Breathes there a man with soul so dead
Who never to himself has said;
'This is my own, my native land?'

———

At Bannockburn in 1314, the Scottish stuffed the English and sent them homeward to think again. But in 1745 they came back and kicked the shit out of us at Culloden Moor in the Battle of Culloden. Bonnie Prince Charlie had arrived on the west coast of Scotland, and he was feeling confident because all these big hairy-arsed Scottish men were waiting for him with beards coming out of their eyes. Looked like they'd each eaten a bear and left the arse hanging out. And they were all riding big Clydesdale horses, eight feet across the back and testicles like frisbees. Wee Bonnie Prince Charlie trotted out on a Shetland pony. This army charged down as far as Derby, then came back a bit demoralised; the whole thing crystallised on Culloden Moor. There were 9,000 opposition forces – twice as many as the other side. It was tragic – a Scottish civil war. The English forces were far better drilled. At that time the Scots fought with a targe, so the English forces attacked them from the side. War was so simple back then. Even during World War One, you'd go in an aeroplane and bomb someone and then they'd do it back to you. Simple tit for tat. Now there's no one with the bomb. It's so smart it knows exactly where to go and how. It arrives in a taxi. 'Knock knock.' 'Who is it?' 'Bombagram!'

The Scottish forces charged at the enemy, but it was a boggy marsh so they must have been completely knackered when they got halfway. After their victory, the English banned people from wearing tartan and playing bagpipes. They even banned the use of the Scottish language. The Scots were subjugated from then on, so the Battle of Culloden occupies an extraordinary place in the Scottish psyche.

But I prefer to dwell on Scottish triumphs. Take the Forth Bridge, a gorgeous-looking piece of engineering that is known and loved by every Scottish person. A cantilever railway bridge that crosses the Firth of Forth, it's Scotland's Eiffel Tower. You can take your Golden Gate Bridge, your Tower Bridge in London and all the others – that's the one for me. What an incredible feat of engineering. As a race, the Scots are clever buggers. Inventors of television, the telephone, chloroform, penicillin, the pneumatic tyre, the adhesive stamp . . . windsurfing . . . A Scotsman even invented the bicycle! Of course, not ALL Scottish people are builders and inventors of extraordinary things. I tried to make the Forth Bridge with my Meccano set, but it was useless. I only had fourteen pieces.

Visitors to the Forth Bridge are often punched for asking where the third bridge is. And there's another favourite among stupid tricks: When you're pulling up at a tollbooth you say: 'How much is that?' The person in the booth might say, 'Fifty pence please,' whereupon you leap out and say 'Sold!' The residents of Dumfries thought it was a joke when Bonnie Prince Charlie rode through the town and took everybody's shoes. But it wasn't. 'Eh you! Give us your shoes!' 'What?' 'Give us your shoes. Don't fucking argue or I'll stab you!' That was in 1745, but people there still hold a grudge.

————

In the West Coast of Scotland there's a pinky-purply haze on everything. People always think it's the heather but it's not. It's on the trees and everything. I need that every so often. Inverness is the capital of the Highlands. 'Inver' means the mouth of the river. It's on the mouth of the River Ness, which flows into Loch Ness. Some people have the cheek to say, 'Is there a Loch Ness monster?' Course there is! Inverness is a nice wee town, but when they introduced a new one-way system I couldn't find my way in. While I was searching around a man spotted me as he was walking out of a pub. 'Hey, Big Yin! . . . OK, Big Man? Fucking all right! Take it easy!' Then he calls to his mate: 'Eh, Willie! Fucking Big Yin, he's here!' His mate comes out. 'So, it is! Yeah! Oh . . . eh, Big Yin!' Sometimes, when people see me in the street they like to tell me jokes. 'Eh, Big Yin! Did you hear the guy who went to the doctor and said, "You've got to help me, Doctor. I've got sore joints, sore head, sore eyes. I cannae sleep!" "Relax," says the doctor. "I'll examine you." He does the examination and then turns to the patient. "I can't find anything wrong with you. I can only imagine it must be the drink." "Oh, don't be embarrassed, Doctor," says the man. "I'll come back tomorrow when you're sober."'

6

THE HAPPY WANDERER

IN ONE OF my concerts in Scotland there was a man wandering up and down the aisles. I said to the audience: 'I know you think this man is going for a pee, but he's a hiker, a rambler. This man should be saluted. This is actually a public right of way here. He does this every fifteen minutes to keep it open for you and me. I think we should give him a round of applause. For far too long he and his kind have gone unrecognised. He has been keeping the highways and byways open and free from vagabonds, thieves and fucking perverts . . . As a matter of fact, I think we should all give him a verse of "The Happy Wanderer".' Then I sang:

'*Oh I love to go a-wandering, Along the mountain track*' and the whole audience joined in with the chorus: '*Fallderee fallderah . . .*'

I love it when things like that happen. That's when it becomes a concert, a real one-off. There won't be a guy wandering around every night. When something like that happens, the audience know they're watching a comedian who's not just a joke-teller, who doesn't just turn up with a bunch of jokes.

A true Rambling Man accepts that he has to think on his feet. Things won't always be tickety-boo. You can be stuck for a place to sleep; plans you thought you'd made might fall through and you'd have nowhere to stay so you'd have to sleep out. I remember being in Aberfoyle near Loch Lomond and having to sleep under a bridge. This bridge had three arches. I went to sleep under the middle one, but I woke up a few hours later and realised the water had risen and my legs were submerged. I had to move my arse. I walked along the nearest street and saw some empty, half-built houses, so I entered one and slept in the rafters. I lay balanced between two planks, so God knows what kept me up because it was only plaster I was lying on. By rights I should have fallen down into the room. But I didn't. I slept peacefully.

I did lots of things like that, just making do with what I could find – shop doorways, empty sheds . . . It was usually good. People were different back then, they didn't really mind – in fact, they looked on it nicely. Hitchhiking was a normal thing, and most people who picked you up were nice to you. That was my experience anyway. 'Where are you going?' 'What are you gonna do there?' Tra la la. It was lovely. They appreciated hitchhiking and it was a great way to get around. People would give me food and all sorts of other stuff – records and things they thought I might like. See, I always had an instrument with me. That was a big sign that maybe I was on my way to play a gig. Sometimes they would put me up for the night. 'Come on . . . you can stay at my house!' The sixties were remarkable. And then we began to hear that people were being attacked on the road. Fucking shame. But before that there was a lovely atmosphere, a joyous feeling. In the sixties, it was all new. There was a free-spiritedness and optimism. It was the perfect decade for Rambling Men because there was a lack of regard for convention and responsibilities.

In those days, homelessness was looked upon a little differently. Homeless people have existed throughout human history, but they are different to rambling folk. You can be a Rambling

Man and not have a permanent address or a home, so technically you're homeless, but not every homeless person is a Rambling Man. Again, it's all about your mindset and your outlook on life. And hitchhikers could be considered homeless – at least while they're on the road, which could be for years. Hitchhiking is a crucial aspect of the Rambling Man's travelling life – or at least it was until they started to pass laws about it, and motorists began to think it was a risky business. There are a couple of states in America, for example, that have outlawed hitchhiking, but quite a few people continue to do it anyway. It's part of their history and culture. During the Vietnam War, a driver with a full car who came across a soldier wanting to hitch a ride would tap the roof of their car to signal that there wasn't enough room.

I've been lucky enough to find room in hundreds of strangers' cars and have spent many great hours shooting the breeze and enjoying the free ride. I used to go on holiday alone, just hitch-hiking with my banjo, and I really enjoyed it. Meeting people on the road, playing music with them. Sometimes I'd spend the night in railway stations, sleeping on a bench in the waiting room. I attached a fishing line between my ankle and my banjo in case anybody tried to steal it. I remember sleeping in one waiting room somewhere where it was bitterly cold, but the heater only stayed on for a few minutes and I had to keep getting up all night to turn it back on.

———

Once, when I was about nineteen and on holiday from the ship-yards, I wanted to go to France. I hitchhiked to Dover from Hamilton in Scotland, and then I had to take the ferry to Calais. I had Joseph Heller's book, *Catch-22*, in my pocket and I was desperate to read it. I read it on the boat, then fell asleep on the cabin floor. I was woken up by the crew, who were taking off back to England. I got off the ferry, but I was on my own and I had nowhere to go. I started hitching again, up to Dunkirk where my cousin was and some of the guys from Glasgow. One guy picked me up and I sat in the back of the car with his daughter. We were all having such a laugh, I was tickling her to make her laugh and he was laughing too. He stopped to get ice cream and he even got me one. So there I was, sitting in the back eating my ice cream with the man's daughter. It was great.

When I got to Dunkirk, I walked through the town to the youth hostel, which is on the coast, and I met my cousin John – he was a Rambling Man. He didn't have a job, so he was just hanging out in Dunkirk. He had arrived at the youth hostel and quite liked the people, so he decided to be there for a while. I'd intended to go to Amsterdam, but I never made it. I just floated around with John and played music. I didn't have a banjo with me, but I had an old guitar tuned like a banjo. We used to play in the street and people would give us money.

I met a lovely woman in Dunkirk named Suzette. She was about eighteen and she was kind of round and buxom. She had curly dark-brown hair and big cheeks. She was eating liver that she had cooked for herself in the hostel, and it smelled delicious.

I was sitting staring at her, and I was starving. She looked up and saw me staring at her. I went: 'Bonjour.' She said, 'Bonjour, monsieur.' I said, 'Would you like to go for a walk? Promenade?' She said, 'Oui,' so we went for a stroll, and I was wishing I could speak French. We saw a sailing boat and I said: 'Bateau!' She said 'Oui, bateau.' Then we walked further, and we saw a fishing boat. I said, 'Poisson bateau' and she said 'Oui. Poisson bateau.' She must have been getting fed up with my line of conversation.

We stepped into a café and there was a glass case with cakes in it. One of them was shaped like a fishing boat so I said, 'Poisson bateau gateau.' And she burst out laughing. She bought me liver and onions. Delicious. Whenever I think of her now, she reminds me of liver. It was a nice time to be alive, the sixties.

———

Hitchhiking is a way of leapfrogging to your destination, but to explore a place you need to wander around in it slowly. My favourite city to walk in is Manhattan. I always find something new there. You meet great people on the streets in New York, selling pretzels and hot dogs. One guy I always passed sold wee kites, another sold toy birds that chirped when you put water in them. 'Get your birds here!' I walked past him once wearing a vintage bowling jacket with the name 'Buster' embroidered on the front. He shouted after me, 'Everybody knows your name now, Buster!' There's only one rule to walking in the city – you have to know a place where you can pee. My solution? If you need to pee, buy a coffee, you stingy bastard. I liked to walk to the music shop in Mannys around 45th Street and 7th Avenue. It sold new and second-hand instruments, and it was wonderfully colourful inside. And there was another music shop I liked on Broadway that sold sheet music, instruction books and records. I also used to love going to the Bottom Line, which was a small music venue where you could see live performances by all kinds of musicians. It's not there any more but it was a legendary club, between Mercer and Greene Streets in Greenwich Village. I saw Loudon Wainwright there once. I played it myself about twenty years ago – I remember Keith Richards came to see me. I liked walking around Union Square, which is near where I used to live. I enjoyed having tea at The Coffee Shop, and I'd go wander into the giant bookshop, Barnes and Noble. I saw Loudon live in Barnes and Noble too; he was doing a wee concert there and promoting a book. I saw John Prine in New York once – I loved him. He was a nice friend to have. Fucking Covid. Took so many brilliant people.

I love meeting real New Yorkers – people who were born and grew up there. They remind me of Glaswegians. They're funny, and they tell it like it is. They don't sugar-coat anything. A Glaswegian friend of mine, Archie, was working on a ship that docked in New York, so he got off to take a walk around. He saw a couple of policemen standing on a corner twirling their sticks the way they used to do, and he went up to them. 'Excuse me . . . can you tell me where the Empire State Building is?' One of the cops said, 'You got a dollar?' 'Yeah,' said Archie, reaching into his pocket. The cop continued, 'Then buy a fucking map!' The New York police officers were always nice to me when I lived there. 'Hey, Connolly!' They didn't want you to stop and speak to them, they just wanted to shout your name. Acknowledge that they know who you are.

———

Edinburgh is another wonderful city to walk around. The whole place is built clinging to the face of an incredibly steep hill and the city streets are just oozing with history. It's easy to trip on the cobblestones, though. When people trip on the street, sometimes they pretend they've seen someone they know. I love seeing that. That kind of thing always made me laugh and I found it made others laugh too. It illustrates something I've heard said about comedy – that it's the ever-present threat of inanimate objects. Well, I suppose it at least applies to slapstick. The Three

Stooges, Charlie Chaplin, Harold Lloyd, Laurel and Hardy special-
ised in that kind of rough-and-tumble act. They used to make
me scream. I see modern versions on American TV in programmes like *America's Funniest Home Videos*, which are just
clips of people having accidents. They're very popular, but sometimes they just look fake or cruel.

When you're in Edinburgh you have a spectacular way of telling
the time. Exactly one o'clock? *Boom.* Cannon on Castle. I went to
the Tattoo once. This confuses foreigners. 'How can you "go to a
tattoo"? Don't you "go to *get one*"?' The thing I like best about
the annual Royal Edinburgh Military Tattoo on the esplanade of
Edinburgh Castle is seeing the guys dancing Scottish reels together
– all those kilts swinging together at the same time. 'But Billy?' I
hear you ask. 'Why's it called a Tattoo?' Buggered if I know.
Someone said it was the sound of tapping because they used to
tattoo you by beating two sticks in a drumming rhythm. I just
looked it up and it was something to do with closing the pubs
early so soldiers would go for their kip. Why have people
throughout history tried to stop men enjoying themselves? My
friend told me he gave his wife a big glass of whisky at Christmas.
She took one sip and spat it out. 'How can you drink that?' He
replied: 'See? You think I'm out enjoying myself every night?'

I used to walk to and from the gig I played at the Usher Hall
in Edinburgh. It's a good old hall, fit for the grand city of
Edinburgh with its beautiful wide, open spaces and stunning
views. There's life in the town and plenty of live music, poetry
and theatre. And Edinburgh has the best comedy festival on
Earth – well, the best I've ever seen – the Edinburgh Fringe
Festival. It has a lovely rough edge to it, an egalitarian mix of
amateurs and up-and-coming professionals. There was a time
when the well-known Traverse Theatre would host a Fringe

promotional event. The Traverse had a little courtyard, and you would bring your poster and sing a couple of songs from your show, and people would decide if they wanted to see it or not. I was performing in something I co-wrote called *The Welly Boot Show*. We were standing in line to do our two songs, and there were a couple of miners behind us, then behind them was Yehudi Menuhin. He had his violin under his jacket because it was raining, and he was trying to keep it dry. I loved that he was standing in line with everybody else. After the miners, he went forward and did his bit and – no surprise – he brought the house down. I thought, 'Yeah, I like this. This is the way it should be.'

——

Walking in a town is one thing but walking through the country-side allows you to be part of nature, in the middle of the landscape of wheat fields, birds, animals, feeling the wind and smelling the road – not just watching it whizz by. A great place to do this is Northern Ireland. You can get lost in the countryside there, it's great. And there are fairy trees. In Celtic folklore, a lone hawthorn tree growing in the middle of a field is a magical gateway or portal between our world and the world of fairies. Sometimes these trees have a circle of stones around them. Farmers never cut them down and will go miles to avoid them in case the fairies get upset. People there hold local lore very, very dear. I'm not one to go saying it's rubbish because it kind of appeals to me. In years gone by, roads were rerouted to avoid the fairy 'thorns' and building plans have had to be changed. If you damage one, it's bad luck. In Fermanagh there was one in a quarry, and they wouldn't move it. But finally they did and, shortly after, there was an explosion that killed three people. What can I tell you?

I know lots of people who believe in fairies. It's a deeply held feeling in Ireland. They're not fairies like we know from children's books who live among the dandelions. They are other-dimension people who came to live among us thousands of years ago and are still here. They live in tunnels and warrens but they're not wee people; they are normal-sized. They come out on various occasions to join in human celebrations and to take part in funerals and life generally. They share their life with the people. We can't see them, though – they are invisible, although people sometimes see them out of the corner of their eye. A belief in fairies is not something that coexists with Catholicism, but many Catholics believe it secretly – just keep it to themselves.

The trees where they live are kept miraculously clean and tidy, but when the branches fall close to the tree they are left there because that's as near as people will go. They take it all very seriously and they don't take jokes about it well, and there's really nothing to laugh at. They'll tell you stories about people who were opening a supermarket who were told they couldn't do it because it was on fairy land but they did it anyway . . . and they lost all their money. They take this as proof of the power of the fairies. I knew a fairy tree at the border of Northern Ireland and Ireland. You knew it immediately you saw it. Wild and beautiful. When you pass by, you have to say: 'Good morning, fairies!'

———

Walking can be hard. I visited Caithness near John O'Groats, where waterfalls go up instead of down. It's an extraordinary place. You get the ultimate aerobic exercise there. There are sheer cliffs leading down to the sea. If you wanted to wash your clothes by the sea, as many people did, it's the ultimate aerobic exercise: a zillion steps up and down. And women would bring their baskets to pick up the catch caught by fishermen. They would race back up, and the first one to the top was awarded a free mackerel. Those women had buns of steel.

I love the north-east of Scotland. It reminds me of my folk-singing days. We often used to play in John O'Groats and Wick. You changed trains at Inverness and then the world changed. The scenery is different. The land is very harsh and hard – almost like tundra. Wick can be brutally cold. When I was there, Malcolm my sound man would turn to me, teeth chattering, and say: 'Billy – remember when we were sitting on the beach in Dubai roasting in the sun, and you said, "It won't be long till we're in Wick?"' That's just another version of the Scottish tendency to see the sun and say, 'We'll pay for this!' In the north-east, the weather changes rapidly – and some people see this as a failing in the place, but personally I see it as Scotland's best attribute. The weather can change within five seconds. You don't know what clothes to wear, and life is generally unpredictable. I just wish the people on TV giving the weather report would tell it like it is: 'It's fucking pishing again. Watch yersel on the mountains: someone fell off, it's fuckin' slippery they are . . .'

Living up there in such tough conditions can be hard. It gives people a special character. I've found they are somewhat distant people who don't communicate readily. But, like all distant people, when they do decide to communicate with you they are warm and welcoming. When you make a friend of the north-eastern people, it's for life. You'd have to have an operation to have them taken off you. Very, very good people to have on your side. I've had many happy years playing there and just wandering around. You see lots of ramblers up there in the Highlands.

Ramblers are not the same as Rambling Men. They are people who go hiking along a specified route, with the intention to see sights, caves, rivers . . . and then they go home. They go walking specifically to spot those kinds of things, whereas Rambling Men just leave without really caring where to go or what might be there. Ramblers have a Rambling Man spirit, but they are organised. They wear anoraks. They're a bit . . . olive green. They like to think of themselves as part of the countryside, but nobody is really, except the farmers. And the farmers are very protective of their land. They don't like people walking on their fields.

———

During my tours, it was always important to me to walk around the town before my gig. I would fly or drive or ride to a town on my bike, but then I'd always go for a walk to learn what kind of place it was. I'd see the people, look in the shop windows, see the town centre, the monuments – and talk to a few people. So, once I stepped onstage, I'd have a good sense of who the audience was, what they liked and what they found funny. Without that I would have been lost so walking was an essential part of my act. You can tell a lot about your audience by looking in shop windows. I once did an interview in Australia in a wellington boot store while I was browsing. 'A lot of people just dismiss the welly as an ordinary piece of rainwear – nonsense,' I told them, and then started to pick up the various boots: 'Here's your worker's boot . . . the see-through flasher's welly . . . your liberal welly – the true blue.' And then I found the best of them all, the Scottish: 'Here you have a low-cut, tartan welly, great for Highland dancing, good sturdy see-through heel for the high kicks, with elastic gussets – say no more.'

I was one of the first people to walk over Sydney Harbour Bridge for fun. The welly is not climbing attire so I did it in my cowboy boots. That *Crocodile Dundee* chap – Paul Hogan – actually worked on the bridge, so he'll have traversed it. But not many people apart from the workers and builders had walked over it at the time, so I was one of the first. I learned the bridge was opened in 1932, and the premier of New South Wales was supposed to cut the ribbon. But a random Irish guy called Francis de Groot charged up on a horse with a sword and sliced the ribbon first. He was carted off to a psychiatric hospital but was found to be sane. They charged him two pounds for the ribbon and four for disturbing the peace. I love the lunatic edge. At the time I got to walk over the bridge I was doing concerts at the Sydney Opera House, and I asked one of the guys who maintains the building: 'How can I see the Opera House from an angle people haven't seen it from before?' He said: 'You can walk on the roof.' So, that same afternoon I was up there looking down from that incredible piece of architecture.

I've always admired extraordinary examples of engineering all over the world, like the Eiffel Tower, the Gateway Arch in St Louis, the Seven Mile Bridge in the Florida Keys. Going across the Forth Bridge in Scotland is a special joy. Especially the old one – the Victorian one. Across from the Victorian one was the modern one. I was once filming there and said: 'That modern one has one function: to stand upon and look at this.' I always marvel at feats of engineering like that. I say to myself: 'Men have put that together with tools.' It honours the glory of working men. The things they've built. I loved drawing attention to them in my TV travel shows. I remember being on a loch in Scotland and showing the camera the man-made islands that conceal Glasgow's water-works. They just look like islands with coniferous trees on them, and they're very beautiful – you'd never know they are functional. But every day the structure inside them pumps millions of gallons of fresh water into the city. I like to highlight what humans are capable of. That's the proud welder in me.

When I went walking in different parts of the world, I used to take photographs of things I liked. I once took a series of crazy pictures during one of my world tours. I photographed the Sydney Harbour Bridge close up – just a few steel girders. When I was in New York I took a picture of the Empire State Building from about eighteen inches away – it just looks like a piece of concrete. And later, when I was in the Sahara Desert, I pointed my camera down at the sand. When I got home with what I thought were very arty World Tour pictures, I showed them to my father. He thought I was off my head. He didn't quite get the joke.

I enjoy wandering around good art galleries and museums. I had special access to the Vatican in Rome, because I was filming there, making a TV show called *The Bigger Picture* about Scottish art for the BBC. I walked around for days and saw fantastic art that has been hidden away from the main displays for centuries. For the first episode I traced the journeys many young Scottish artists made to Rome in the eighteenth century. In the third episode, I even got to have my portrait painted by the famous Scottish artist John Bellany. I thought he might have given it to me, but he didn't. Bastard.

———

Whenever I'm out walking, I have to keep my eye out because I can be caught out. I was out one day in London and a guy came running up behind me. 'Billy! Billy! Could you sign an autograph for wee Gavin my grandson?' 'Sure.' A wee boy came up beside us. I wrote on his wee paper: *To Gavin – Best Wishes, Billy Connolly*. Then I started walking away, and I heard: 'Billy! . . . Billy! Sorry again to disturb you on your walk,' he says. 'But Gavin was wondering if you could tell him to fuck off.' 'Sure,' I said. 'Gavin? Fuck off!' The wee boy went 'YESSS!'

I walked a lot in Rotorua, New Zealand, and around Te Puia's Whakarewarewa Valley. It's known for its geothermal activity, so the lakes and bubbling mud pools there look like your porridge is ready. The forests are ancient and lovely to hike in, and the town is bright and nice. And they have an extremely virile geyser that has many daily ejaculations (some prefer the word 'eruptions', but if you saw this thing you'd agree with me), shooting thirty metres in the air each time. But the highlight of my New Zealand tour was walking via the Ninety Mile Beach to where the Tasman Sea meets the Pacific Ocean, in the northernmost part of New Zealand. At the tip of the Aupouri Peninsula lies Cape Reinga, or Te Rerenga Wairua – which is a Māori name for the spiritual pathway. It's a very special, sacred place. The Māori people believe it is from there that the spirit leaves for the ancestral home – the end point of their earthly world.

———

Two guys were talking in a pub.

'What would you do if we just found out we had three minutes to the end of the world?'

'I'd shag everything that moved. What would you do?'

'I'd stand perfectly still.'

7

WITCHETTY GRUBS AND CULLEN SKINK

—

MOST RAMBLING MEN can't afford to be too picky when it comes to food. On the road, you eat what's available. Some of them can cook on an open fire but, even when I used to go camping, that was a bit beyond me. I didn't do much of that. I did cook potatoes though. I would wrap them in tinfoil and throw them on the fire. They were delicious. And if I caught a wee fishy, I'd cook it in foil too. I've been living in Florida for several years now, and people here love to cook outdoors. Not me. I mean, I like a burnt sausage as much as the next guy, but I prefer to cook a wee curry in my air-conditioned kitchen without competing with insects, birds, palm rats and uninvited humans. Mind you, when I breathe in the smell of barbecued meat that comes wafting over my neighbour's fence I think: 'Maybe I'm missing out on something great!' And the smell of the sweet, spicy food the Haitian family down the road cooks in their backyard really makes me ravenous. They sit around playing music and eating, everybody laughing and shouting . . . I wish they'd invite me. In recent years at Christmas, the men in our family have established a tradition of deep-frying a turkey – which is most definitely an outdoor activity, according to the news that always pops up about the kitchen explosions and house fires they have every year at turkey time.

I grew up on school dinners, so that's the kind of food I like best. Shepherd's pie, sponge and custard, mince and potatoes, sausages, apple pie and custard, caramel pie and custard. The only things I didn't like were the horrible vegetables all at the side of your plate.

But I would have walked to the moon to get more shepherd's pie. Now, my own macaroni and cheese is next to none, but I could eat curry every single day. When I'm on the road, though, I try to be open to local cuisine, and I'm rarely disappointed. What's the point of going somewhere exotic if you're always going to seek familiar things? If you always stick to your home favourites, you'll miss out on some of the tastiest dishes in the world, such as those you can find at the roadside food markets in Hong Kong, the yakitori bars in Japan, and the Cuban food trucks in the Florida Keys.

———

I am very fond of American roadside food, although I wouldn't eat it all the time, because it's a sad fact that it's not exactly good for your colon; but it's perfection when you're on the road. Traditional American diners are great for breakfasts. My favourite is bacon and eggs accompanied by American biscuits with gravy. I love the way Americans do bacon – it's a very different cut to British bacon, which is wider at one end and usually has a lining of fat down one side. The American bacon is a narrow strip streaked with fat that becomes crispy when cooked. It's delicious. My fried eggs have to be 'over easy'. It's great to sit down to a breakfast like that. And if you're in the mood, home fries are wonderful, but they really fill you up. If you're driving on your motorbike in the rain, it's great fortification. But I don't think American coffee matches the breakfasts. Tea is much better. It's in the same gang. Some greetin'-faced British people don't like the way Americans make tea. It's true that some essential mistakes are usually made – a mug of lukewarm water with a wee teabag in a paper bag sitting beside it – but that's not going to change, so it's best to just shut up and get on with it.

Some roadside diners in America specialise in weird stuff, like alligator jerky and roadkill. I like to try those kinds of delicacies, but I think very little of that kind of thing goes a long way. I mean, wild dogs may be tasty, but you can't beat a nice bit of rat in the morning . . . on second thoughts, just give me a hot dog with some fries. My typical lunch on Route 66 was a double hamburger with a cheeseburger and a cup of tea. Dinner was easy – we would go to Chinese restaurants and I'd get prawn fried rice with Chinese tea.

In Texas I liked to eat the chicken fried steak. That's steak that's fried in batter the way chicken often is, and it's great. And I drank soft drinks like the locals. For breakfast I liked the Mexican dish huevos rancheros. Texas is great for that. At first, I thought huevos rancheros was a cereal called Ranch Cheerios, but it's fried eggs and hot sauce. Tasty. I learned it was a great favourite of one of my heroes, Peter Cook.

I enjoy good soul food when I can get it. I tried crawfish pot pie in Louisiana, and it was wonderful. The crawfish was minced up and tasted a bit like prawns but even more delicious. Hank Williams sang about the bayou and crawfish pie: *Me oh my, crawfish pie* . . . I like gumbo too – strong, thick soup made with shellfish or meat, celery, onions and peppers. I love it when you get whole shrimp in it.

But my favourite so far has been at Sweetie Pie's Restaurant in St Louis. The proprietor, Miss Robbie Montgomery, used to be one of the Ikettes, the legendary backing singers for Ike and Tina Turner who would tear up the stage with their electric performances. 'Yeah, I used to shake my tail feathers,' she laughed, 'but now the tail feathers broke, I can't shake it any more – I've gotten too old for that!' She told me that when she was touring with the Turners she would cook for herself and the rest of the performers, as African Americans were not welcome in many restaurants. 'Everything was segregated back then, so there wasn't a lot of places for Black people to eat so we had electric skillets, and we would get in the hotel rooms and cook,' she said. Miss Robbie was also a backing vocalist for other musicians, like Stevie Wonder and Barbra Streisand. Now she is in her seventies and famous for her soul food, which some consider to be the best in Missouri. I tried her collard greens (the best vegetables I've ever eaten) and her mac and cheese – real Mississippi-style cooking – and boasted to Miss Robbie that I cook mac and cheese myself. She was very gracious. 'Well,' she said, 'I don't want to beat you. Just put me in the race!'

———

Australian food is unique. I tried kangaroo curry in a Sydney restaurant. It wasn't all that good. I didn't like the texture. Foster's lager and Vegemite are food of the gods though, in my opinion. But the 'bush tucker' or bush food is a whole new level of haute cuisine. In the dreamtime culture of First Nation Tiwi people, civilisation began on a beautiful island with white sand in the far north of Australia, where you can catch and eat mud crabs, or 'muddies'. I met a fabulous Rambling Woman called Eleanor who taught me how to catch muddies. You use a stick with a barb on one end to hook the crab and pull it out by its shell. Eleanor caught another dozen mud crabs and a big blue swimmer too. 'That's your lunch,' she said to me, which was when I asked her to marry me. I said, 'I need to be married to a woman who can catch muddies in the morning.' She showed me a trick to stop the crab from nipping you with its claws – she snapped two of its back legs off and jammed one in each claw to prevent it from closing them. I felt sorry for them but they were tasty wee buggers. Ever had the other kind of crabs? I've had boiling piss, I've had the whole shootin' match. Fucking Agghhh!!! But nothing compares to crabs. When you look at your pubic hair and it's become an adventure playground . . . Fuuuuuck!

There was no end to Eleanor's talents. She cracked open a tree to find a worm called a witchetty grub that's considered a real delicacy there. It's actually the large, white larva of a moth. She pulled a couple out immediately. They were quite long, and they were wriggling around. I ate them both live, because that's the kind of devil-may-care adventurous savage I am. A witchetty grub tastes a little bit like an oyster; it's slippery and revolting. But when Eleanor passed me another, I ate it. I've had worse things in my mouth. Bush tucker is quite in vogue. A lot of people want to try it, and not just the poor wretches stuck on a jungle TV show. But I don't understand why it's called 'bush'. Shouldn't it be 'The bushes?' For God's sake. It's like calling the Sahara 'one fucking grain of sand'. But 'the bush'? 'Oh, this must be the bush they talk about. Four million kangaroos hiding behind it.' Shake the bush and see what happens . . . *baoing baoing*, they all jump out . . .

Freud says whenever living creatures meet, they immediately say: Can I eat it? Can I shag it? all in a zippidy-doo microsecond. Or – can I shag it while I'm eating it? Well, I made a BBQ in the bush, but it wasn't your usual backyard menu. It was frill-necked lizard and leg of magpie goose. How fresh would you like your goose? That thing was flying this morning. Call me old-fashioned, but if I met those animals in the bush, I'm not going to think either 'eat' or 'shag'.

Eleanor had vast knowledge. She showed me a giant seed pod – a kurrajong. They roast it in the ground, and it tastes like popcorn. And she knew how to make baskets out of large leaves, use a soap plant for washing, and plant green plum that tastes like ginger beer. For First Nations people, the bush is like a free supermarket. Shops everywhere but everything is free and it's open all hours. When we were walking along the beach to take the boat back to the mainland, some of the film crew jumped in the water to swim, but shortly afterwards they came running out because one of them saw a shark. 'Shark! Shaaarkk!' he called, flailing about in the waves and overbalancing in the soft sand as he tried to get to the beach. But Eleanor and the other local women had a completely different attitude. When they heard him yelling 'Shark!' they immediately grabbed an axe and ran into the water. 'Shark? Where? DINNER!!!'

But it wasn't only Eleanor's foraging skills that impressed me. I was really affected by her because she was probably the warmest person I've ever crossed paths with. My daughter Cara was travelling with me at the time, and when I pointed her out, standing with the film crew, Eleanor went crazy. She immediately ran to her, gave her an enormous hug and made a huge fuss of her in such a loving way. Then later, when we were on the boat returning to the dock, Eleanor pointed out her son, who was waiting for her on shore. 'Look!' She was running about telling everybody. 'See him? That's my boy! My boy!' She was so proud of him. That kind of love, demonstrative motherly love, is so alien to me. It makes me feel a wee bit uncomfortable . . . and touches me deeply at the same time. Eleanor impressed me terribly, and she taught me a lot.

———

The town of Alice Springs is at the very centre of Australia, and there's fuck all else around it for many, many miles, so you have to look after yourself without a microwave. That's where I was shown how to mix a damper – that's a swagman's traditional bread made from flour, butter, salt and water. If I'd only known about the damper when I was camping in Scotland, I'd have been in great shape. It's such an easy, delicious meal with very few ingredients. The traditional Australian swagman is a real Rambling Man. Does it right. He carries his swag – a rolled-up sleeping blanket. Australians used to make a lovely sleeping bag where the outside part is oilskin. It has a hood that comes up and is held out away from your face; I always wanted one of those. They really do the ruffty-tuffty thing very, very well, Australians. The damper turned out well, and the whole process was quite exciting. I thought I'd burned it to a crisp because it was all black on the outside, but when I opened the charred bit it was excellent. I have a cookbook at home with a recipe for damper. It's also got a recipe for making a purse out of a red kangaroo scrotum. Yum.

At my bush campfire I also made Billy Tea – a traditional tea made in a can. I gathered the wood and lit the fire – I was a Boy Scout so I knew how to do all that from my camping days. I made the tea in the same kind of can we used in the shipyards – the one that powdered baby milk comes in. It wasn't until your can was black that your tea tasted great. Tea made in a can is phenomenal tea. You just drink it black with sugar. After that, having tea in a cup was always a dead loss. With my bush tea and all, the whole meal turned out to be brilliant. Then, in true Rambling Man fashion, I sat under a tree and played my auto-harp. It makes a lovely noise outdoors.

You can tell a lot about a nation by the food. A great delicacy in Australia is the pie floater. It's proper food. None of yer continental rubbish. It's basically a meat pie with pea soup all over it. You put brown sauce on it and vinegar. Somebody in Australia holds the record for eating pie floaters – nine in three and a half hours. I reckon I could do that. I heard Joe Cocker always had shepherd's pie in his dressing room – except in Adelaide, when he had a pie floater.

I'm very fond of Australian sweeties. Minties are my favourite, although they suck your fillings out. I like Violet Crumble bars too (I call them 'Violent Crumbles') and Cherry Ripes. In the cake department, I especially love lamingtons. The lamington is very special to me. It's my New York pizza. I don't feel I'm in New York till I have a pizza and a Coke, and I don't feel I'm in Australia till I've had a lammy and a cup of tea. But you can't beat pies and, personally, I will always say 'Aye' to a pie.

Barry Humphries wrote a poem that went:

> *I think that I could never spy,*
> *A poem as lovely as a pie,*
> *A banquet in a single course*
> *Blushing with rich tomato sauce.*

That's Barry. I miss him.

I once exchanged poetry with a man over a large tamarindo – which is a tart, sweet, tasty Mexican drink. It was in Miami, during the filming of my *Ultimate World Tour* series. In the Miami public library, you can not only borrow a book of poems from the library, you can borrow an actual poet, so I checked out Michael A. Martín. We went to a restaurant named Yambo's. When we sat down, Michael said, 'You checked me out, so I'm going to read you a poem now. It's part of the job . . .' And off he went with a poem by Campbell McGrath:

> 'The Key Lime'
> *Curiously yellow hand-grenade*
> *of flavor; Molotov cocktail*
> *for a revolution against the bland.*

———

'That's great,' I said, 'I could give you one of mine? It's called, "I'd rather be a sausage" by Billy Connolly':

> *I'd rather be a sausage*
> *Than a British man of war*
> *Or a caterpillar with a broken arm.*
> *Corduroy braces are all very well*
> *And give no immediate cause for alarm,*
> *But the sausage is a mighty beast*
> *Who serves only to please,*
> *In fact, he is the mightiest there is.*
> *Content to lie in frying pans*
> *For hours at a stretch,*
> *Singing sizzle sizzle sizzle sizzle siz.*

———

I made that up about ten years ago when I was scared on an aeroplane and needed to preoccupy myself.

In most places I've been, I've rarely been disappointed when I've tried the local food. I went to Oslo for the Nobel Prize ceremony. Nelson Mandela and F. W. de Klerk were getting the Nobel Peace Prize jointly that year, so I was well pleased to have been invited. My wife wanted to meet Mandela, so she came with me. We were eating dinner at the event, and it was reindeer. 'Oh, give us a bit of that!' The old Rudolph and chips here. So, I'm eating my reindeer and I'm wishing I had a glacé cherry with me so I could say 'I've got the nose! I've got the nose!' Really fuck everybody up. The room was half ANC, and you don't get too many reindeer running round Soweto most mornings. So, one half of the room was describing reindeer to the other half of the room, and they were all making antler shapes with their fingers on top of their heads . . . I collapsed with laughter. You know the way your legs go? I laughed so hard I snorted and reindeer came down my nose.

As I've already mentioned, I don't like to eat outside. There always seems to be wee annoying things buzzing around trying to nibble at me or my food. But I make an exception when it comes to two special beach restaurants – Doyles, a wonderful fish restaurant at Watsons Bay in Sydney, and Mgarr ix-Xini in Gozo, Malta, where my pals Noel and Sandra serve the freshest, best-cooked fish I've found anywhere. Mgarr ix-Xini is a brilliant place where you can sit on your arse under a shady tree and while away the afternoon while your wife goes diving, swims around to the next bay, or climbs a sheer cliff and jumps off. Not to be outdone, I get my own exercise – getting up to order another Cornetto.

Pamela is half woman, half fish. There are Irish legends about silkies – seals who can shed their skin and become human women at will – and I think she may be one of them. She is Australian, but she was born in New Zealand. When I first went to New Zealand it was difficult to get anything to eat after a gig. That tainted my view of the place. I know that's hardly the right way to think about a whole country, but when you're on tour in a place you need basic comforts above all. During the day, me and the crew would see a restaurant we fancied, so we would go in to enquire about an evening meal. But they'd say: 'We finish at eight o'clock.' We would say: 'But we start at eight!' I didn't get offstage until at least half past ten. They'd just shrug their shoulders. 'Hard luck.' It was like that for my first few tours there. If we were lucky, the restaurant in the hotel would leave us a salad, but it was the last thing you wanted after three hours onstage. And it was like your granny's salad – two lettuce leaves, two slices of tomato, a boiled egg and a bit of ham.

But all that has changed radically over the years I've been going to New Zealand, because enterprising people took over the operation. A new generation has taken over. Eventually we were able to go for Indian food after the gig – my favourite. I remember there was one woman in charge of an Indian restaurant in Auckland. She served everybody and cleaned up. Not only was she the cook, and the waiter, but she did the whole thing. It was so impressive that you wanted to give her a standing ovation when you'd finished your dinner.

———

Māori people traditionally eat eel, but I've never tasted it the way they prepare it. I've never been very fond of eel; I tried it once in London and I was horrified. I hate wobbly food. On a beach at the edge of the Tasman Sea, I once went to dig for Tua Tuas – which are a sort of clam – to make soup and fritters. You do the Twist in the sand and your bare feet will feel one of the shellfish, so you can pick it up and throw it in your bag. New Zealand has fabulous shellfish. My promoter Ian McGann had an aunt who lived in Invercargill and she sent us a Maxwell House jar full of oysters and green-lip mussels. Hurrah! I also like the big clabby doo mussels you get on the west coast of Ireland. Fabulous. The best oysters I ever tasted were in Stewart Island in New Zealand. In the seventeenth century it was the poor people who drank claret and ate oysters – I learned that when I visited the Mary King's Close underground street in Edinburgh. It lies directly beneath the City Chambers and it has a terrible history; it was sealed off because of the plague and they just let people die in there. Awful.

Some Scottish dishes are addictive. I'll go miles to get a good cullen skink, a soup made from smoked fish and potatoes. And I am very partial to the smokie from Arbroath. That's a smoked haddock and it's brilliant. But I'm really a man of simple tastes. Of an afternoon I like to partake of a wee fairy cake and a cup of tea. Although, I like a wee cappuccino as well. But I have to think 'Al Pacino' cos I keep forgetting the name of the fucking thing. Once I even asked for a Robert De Niro. Baffled the waitress completely.

I always asked my promoters to put absolutely nothing in my dressing room – just a cup of tea – but they often forgot. I'd get big piles of sweeties and lots of booze. I enjoyed the sweeties, though. In fact, I'm a fan of the Mars Bar diet. It really works if you do it correctly. Some people make the mistake of eating the Mars Bar, but that's wrong. You don't eat it. You stick it up your arse and let a Rottweiler chase you home. On a regular basis.

My audience tended to munch sweeties during my concerts. When it got too annoying, I'd single someone out and say: 'Did I hear sweeties being opened? Everything that moves in here is MINE. See, you're laughing with a sweetie in your mouth that's what happens. And it shot up your nostril. Oh! – Coke too! You're gonna end up a big fat person.' Then someone would throw me a sweetie. 'Chocolate eclair! My favourite!'

I love a blueberry muffin. Definitely not bran – that's for people trying to shit themselves thin. My doctor told me eating grapefruit burns more calories than the grapefruit has. I remember my answer distinctly: 'So fucking what?' And brown bread is an abomination. As far as I'm concerned, brown bread is a fucking frisbee. They'll say, 'You'll live longer,' and you will. You'll live another fortnight. But it won't be the fortnight when you're thirty-five and shagging like a stag. You'll get the fortnight when you're pissing your trousers and being fed out of a blender and wishing you were fucking dead – THAT fortnight. You smell of piss, but you've got two weeks to go. 'Oh, for fuck's sake have mercy . . . hit me with something. Throw me off the fucking building!' 'No! Brown bread for you!' 'Have mercy! I'm fucking ninety-seven!'

I'm not picky, but here's a few things I hate: I can't get beyond the smell of tripe. There used to be a barrow on my way to school. A woman with a wheelbarrow who used to sell tripe to the tenements. I hated the smell. And I would vomit if someone put Brussels sprouts on my plate. I find it difficult to eat any kind of vegetable, although I can tolerate most of them raw. I also hate revolving restaurants. They're for people who like to spew after their dinner.

But I love Indian food. When I was filming in India, we used to have lunch in a deserted house with no roof. The actors and production team would eat on the first floor where there was a great variety of food, and the crew ate on the ground floor, and they just got dhal and chapatis. I was going in for my lunch one day and I thought, 'Dhal and chapatis? – Say no more.' I went in and sat with the crew having a laugh and ate my dhal. When we were leaving, the manager of the film said: 'Have you ever been here before?' He was amazed at how easily I ate with the crew, cross-legged on the floor. I said: 'No. I've never been here before.' He said: 'I think you've been here before.' Gave me a wee chill. But I learned more recently that my great-great-great-granny Matilda was Indian. Maybe that's why I respond so positively to all the colours and the smells of all the spices wafting around all the time. And to the interesting Indian perfumes that are quite different to what Europeans wear.

I like to cook Indian food myself, inspired by Madhur Jaffrey. I fucking *love* Madhur Jaffrey. I'd like to marry her too. I wonder if Eleanor would be open to a wee *ménage à trois*? My children eat Indian food now because they are adults, but earlier on they wouldn't. There's nothing like rejection of your food by a child. Adults will say: 'You know I haven't felt very well all day. I think I'll go outside and have a glass of water. Actually, do you have any paracetamol or anything?' But a child will go '*Bleughhhh*!! No!! *Dodo*!!!' When my middle daughter Amy did this to me, I went to her older sister Daisy and said: 'What's she saying?' Daisy had the answer: 'She wants McDonalds.'

If I was on death row my last meal would be fish and chips. Great food. It's a joy. Haddock, Glasgow style. Food of kings. Yeah, and I'd want to wash it down with a pint of heavy. Served by Eleanor.

Travelling with a musical
instrument always helps
you hitch a ride . . .
Posing during my Big
Wee Tour of Britain, '64.

Watch where you put your hands!
Motorcycles always made me more
attractive to women. Took the
bad look off my acne. Kawasaki
Motorcycle Store, Newcastle.

Champagne shower after the
Glasgow to Inverness charity
bicycle run. Not sure why my
mouth is closed.

Is that Nessie?? Monster-spotting on Loch Ness.

Is that your wife's bum sticking out there? Filming in Rome for *The Bigger Picture*, a series about the history of Scottish art.

In the penguin pen at an aquarium in Auckland, New Zealand. With Fatboy the penguin who sang me a wee song.

A wee pick on
my banjo aboard
the *Bounty* on
Sydney Harbour.

I just needed a wee rest:
climbing the Sydney
Harbour Bridge.

Made it!

Big bitey things threatening Pamela and Daisy in Cairns, Australia.

A nice wee catch! Newcastle, Australia.

Quokka spotting: cycling in Rottnest Island.

The mud crab crew: the wonderful Eleanor is by my left side.

See how I eagerly rushed to assist the man digging bravely for vicious mud crabs that can snap your fingers in half!

In Australia with Steve Brown, my manager of many years.

Spot the prick in the desert boots.
Attempting to play Elephant Polo in Nepal.

If there'd been
elephants in
Scotland we would
have invented an
even sillier game:
Elephant Golf.

Through the dreich and the
drizzle – the show must go
on! Filming my World Tour
of New Zealand.

Turkey-hunting twins, Carolyn and Cheri, in Fanning, Missouri.

The original 'balls of steel' in Payson, Arizona.

But how do you really feel about me? Animal sanctuary along Route 66 in Valentine, Arizona.

Son of a gun, we'll have big fun on the bayou . . . Travelling by riverboat along the Mississippi.

Do my eyeballs look the same from the outside? With Farmer Tom at his Marijuana Farm near Oregon, USA.

My hero, Wiley: a yodelling cowboy at a train station in Shelby.

8

SIZE MEANS VERY LITTLE WHEN THE LIGHTS GO OUT

———

I NEVER PREPARED for a tour. I never rehearsed. I worried I'd forget everything I'd ever done in my life and the more I worried about it the more of my show I'd forget. I couldn't remember a word of it. I had this fear that the next audience would be a crowd of Martians who'd hate what I did – which is ridiculous. They were my fans, and they were coming to see me. That anxiety always lasted right up to when my manager made the announcement: 'Ladies and gentlemen – Billy Connolly.' On the words 'Billy Connolly' I would walk on, and all that fear would fall away. I'd have no more nerves. Well, it was too late for nerves at that point, anyway. I can't explain it, but I can tell you that – paradoxically – I loved it. It's the thing I have probably loved most in my life.

In some ways, I considered the relaxation time between gigs to be more important than the gigs themselves. What you do in between concerts can make or break your show. In the early days I used to get drunk and stay up all night, be wild, crash around and have a great time generally, and it didn't seem to affect my gigs as far as I was concerned. But now, when I look at the tapes of those gigs, it certainly did – although it had a certain cachet, a certain charm to it. I gained a reputation for unruliness, and people would flock to see this hairy nutcase. My roadies and I would dedicate each tour to a different kind of alcohol. There was the 'Whisky Tour' and the 'Gin Tour' and the 'Brandy Tour', and the rules regarding limitations on other drinks were strictly adhered to. Is it any wonder I remember very little about my nights of fun and debauchery? I wish I could say truthfully, 'Oh, how I regret those nights of shame!' but I don't; I can't remember them. The pattern of partying heavily after each show at the time seemed normal and good fun – and everyone around me seemed to be doing the same thing – but I did not realise my alcohol use was getting well out of control. By the time I met Pamela in 1979 I had become a fully fledged alcoholic who had frequent blackouts and even violent episodes. Fortunately, with Pamela's help I managed to stop drinking while it was still my idea. I've been sober now for forty years.

In my early folk days – especially when I had just formed the Humblebums with Tam Harvey – we didn't really have much of a choice but to misbehave and live on our wits. Once we played a folk club in Arbroath. We weren't booked to play there – we just turned up and they let us perform. We went down very well, and afterwards the guy who owned the place said: 'That was great . . . but we don't have any money to pay you. If you can come back next week, we'll give you the head-liner spot and you can have the big-time money.' So we decided to stay in Arbroath, but we didn't have any funds for digs, so we just went and chatted up women so we could stay with them, and women who worked at hotels let us stay in rooms without the manager knowing about it. It was easy to get the women to put us up. We were hairy people. Attractive. I'm not going to make excuses for that kind of behaviour, except to say it was typical for the sixties Rambling Man, and I'm pretty sure it was fun for everyone involved. Tam and I played the top of the bill the following week and made some serious dough, so we could afford to eat. Best fish and chips in the world in Arbroath.

———

I've played many, many venues all over the world – some famous, some not, some huge, some wee. Some were exactly the right size, style and shape for their location, while some seemed out of place. Even after I had started selling out big arenas, I would sometimes book a tour of little seaside towns. In 1994 I did a tour of wee places in Scotland. It had been a while since I'd gone back and played there, and I loved it – revisiting the Isles of Bute, Cumrae and Arran. It's the Scotland people sing about, the bits I love. Most years I try to get to the folk festival in Arran. It's brilliant.

That wasn't the only time over the years when I have chosen to do a tour consisting of small gigs, although people didn't expect me to do them at that stage in my career. They would say: 'It's very kind of you to do that . . .' But in actual fact, the small places tend to frighten the life out of me. With a big audience, the applause and the laughter hit me like a big wave, so while they're applauding and laughing you can think of the next thing you want to say. But when it's only 200 people they don't laugh for very long, so it just goes 'Hah hah', and then it's time to talk again. You get caught short all the time. And in these little halls it's like singing to your aunties and uncles, so they're not very impressed. I like performing in small towns, but sometimes the audiences behave as though they don't like you – until the very end of the concert. They are very reticent to applaud and don't laugh very much, but then they go crazy at the end. When you announce that you're leaving the stage, they all cheer and stamp their feet and you think, 'My God – why didn't you do this an hour ago? I thought I was dying the death.' But it's just their way of responding. The living is hard. They come in with the attitude: 'Okay – this is entertainment – entertain me!'

The places I played when I was a folk singer were always small. Edinburgh had a pub with a folk club upstairs that was a legendary place. One year during the Edinburgh Fringe Festival, two guys and I arrived, but there were no slots for us to play in the club, so we just got our instruments and went downstairs and played in the public bar. People loved it – they were all singing and stomping like they'd never heard bluegrass before. When we were finished, I said, 'I'll just nip upstairs to see if the folk club's still open.' Maybe it'll be a bit empty, and they'll put us on. It was empty. In fact, it was deserted – not a person, not a barman around – so, I nipped in. There was a bottle of whisky lying there, so I took it downstairs. It was closing time, so I lifted my banjo to leave, and the barman said, 'Just a minute . . .' I thought, 'Oh no, I've been caught.' He said, 'You were very good. Here!' Gave us a bottle of whisky and a dozen cans of beer. I was so embarrassed, I put the other bottle back upstairs.

When I was a folk singer, I was part of a community of hairy crazies – rambling men and women who more or less lived on their wits. But once I became a comedian, I had to adjust to life on the road and try to behave in a professional, serious way. It wasn't easy. I'd never stayed in a hotel before. I'd never had room service – I didn't even know you did things like that. Even now, I prefer an ordinary hotel room – not a suite. At first, I was taken aback by the life of luxury. Big fluffy towels . . . so big and fluffy you can hardly get them in your case.

———

I built my act in an informal way – I never wrote anything down, except a few headings to remind me to tell this story or that one. And I was unusual, being a hairy Glaswegian guy. The comedians I admired – like Max Wall and Frankie Howerd – were not like me. When we listened to the radio or watched TV the popular comedians were Charlie Chester, Jimmy Wheeler and Dave King – and they were all Englishmen. But when I went to the theatre in Glasgow and saw Jimmy Logan being extremely funny in my accent, singing parodies, funny songs about things I knew – the coalman and the journeyman, the street I walked along every day – it had a profound effect on me. I thought maybe I could take a leaf out of his book. I decided then I would like to be a comedian. And there were other Scottish comedians who inspired me, like Jack Radcliffe and Chic Murray. Chic used to say: 'I was at the Olympic Games and a man came along carrying a big thing on his shoulder. I said, "Are you a pole vaulter?" He said, "No – I'm a German. How'd you know my name was Walter?"' That kind of nonsense really makes me laugh. They should have a statue to Chic in Edinburgh. Never mind those Roman fuckers – nobody knows who the fuck they were, anyway. Put one up of Chic!

Most of what I did onstage was spontaneous. It just came out of my head – which was terrifying because I had to more or less repeat it every night. Somebody once told me that after Tony Hancock had just done the famous 'blood donor' sketch he was in his dressing room, really depressed, moaning: 'Ohhh! How am I gonna follow that?' I understand that perfectly. The way I operate, you have to follow yourself every night. When you do really well you go to bed and you say, 'Oh God, I have to do it tomorrow again.' And the next night and the next – in my case, it could be fifty-four nights in a row.

Even the *thought* of touring makes me very nervous, which is actually a good thing. It helps. The worst is to go on feeling completely confident. To get the best out of yourself you should always be jumpy. Comedians and boxers are very, very alike. We need to be keyed up before taking on the fight. If the boxer comes off alone, he's done well. And if the comedian comes off with laughter still ringing in his ears, then he's done well. I used to come on a bit like a boxer and stride around the stage, with ideas going round in my head, putting out little feelers – to see what they're like, what they found funny, how far I could push the little adventurous boat out. And once it's flowing, once it settles, it achieves a certain rhythm. Mind you, a boxer can train with a sparring partner, but comedians can't rehearse in the same way. You can't just try out comedy in an empty theatre; you have to go onstage with an audience cold. I used to have long periods off and then work intensively for a while, but during the long periods off I felt like I forgot what I did. I forgot what you pressed to make it work. So basically, I relied completely on just walking onto the stage, sensing the atmosphere and getting on with it.

———

I remember the first time I toured Australia with the new promoter I'd just met, a man called Kevin Ritchie. He was driving me to my first Australian gig. I was sitting in the back of the car with my roadie Jamie Wark, and I said: 'Jamie, what do I do? How do I start the show?' He said: 'I can't remember.' I said: 'Try to remember.' He said: 'You do a thing about such and such . . .' I said: 'Oh yeah. What do I say about it?' He told me. I said, 'Oh yeah, that's quite good. Anything else good?' He said: 'I can't remember . . .' and I was about to go onstage in twenty minutes. Listening to all this, Kevin was almost having a heart attack. He was thinking: 'What the fuck is this?' Here we were about to play Townsville and I didn't know my act. But I wasn't worried. I was just casually looking out the window watching mangos falling off the trees – and that made him even more worried. But the concert was a huge success. When I came offstage, Kevin was waiting. He was sweating profusely. He said: 'Do you do this every night?' 'Yes.' 'Jesus, it's going to be a long tour.' Eventually, he accepted that I was different from other performers – and he even started to enjoy the chaos of it, because we had fun.

The energy level needed for my concert was high, but I couldn't summon it at will. It just turned up, seemingly appearing from nowhere. Sometimes I'd been feeling terrible during the afternoon or heard some bad news, or run into a problem getting to the gig, but when I got to the theatre and the lights went down something inexplicable would happen. I would be standing on the side of the stage waiting to go on and there'd be a wee circle of light on my chair, and suddenly I'd feel a surge of electricity as though somebody had stuck a battery in my neck. They'd say my name and I'd walk out into a different world altogether. As a matter of fact, that creativity didn't happen for me in a room – I've never written a word of it down. It happened onstage in front of the audience. I am much more inventive *during* a show than when I am offstage.

A lot of my energy came from anger. I became very jumpy and speedy onstage. I'd get angry at politicians, angry at selfishness, and I'd get angry at pretentiousness. Truthfully, I just love going on the attack. One of the great joys of life is to be handed a microphone, hit with a spotlight and be allowed to say anything you damn well please. Unless you've done it, you can have no idea how brilliant that is. There's no drug on Earth that can equal it.

When I first toured Australia, they were unsure of me there. The press went bananas about my toilet humour. They said, 'His humour is at the level of the lavatory.' So, I held a press conference in a toilet, and they went to town on that. But other people thought it was funny, and my popularity grew. Early on in Queensland my audience mainly consisted of expatriate Scottish guys. It was predictable that some Rangers supporters were going to make it rough on me, and they did. The concerts were working fine, but some of them were just determined to destroy it. Once, at the Guildhall in Brisbane, a guy leapt up onstage and walloped me. He was a fucking idiot – a Scottish-Australian prison officer, which is a deadly cocktail. He was a dedicated non-thinker. I think he learned to walk upright that morning. He said, 'My wife's ears are not garbage cans!' He smacked me right in my face and I fell on my arse. I felt so stupid. There was a bit of a stramash, and the police were called and the whole thing was a nightmare. I can't remember too much about it. I've lied so much about it over the years and exaggerated so much out of all proportion that I really don't know what the truth is – probably lost somewhere in the middle of all the fun, lies and exaggerations, which happens to me quite a lot. A made-up story can become the truth to me after a wee while, or the truth gets conveniently lost. I hope I'm never involved in any huge court case because I'll be one of those guys who sits in the witness box going: 'I do not recall.' Yeah, my real life and my act become intermingled so much I have a hard time separating them.

At the time when that guy came up to hit me, I was telling a great joke: A guy fell off a cliff and he was dangling at the end of some kind of tree branch he'd grabbed as he was falling. Below him there was a sheer drop to certain death. He was terrified, and he cried out: 'Is anybody out there?'

And a voice said: 'I'm up here. I'm always up here, my son. Don't panic. I'll look after you.'

'God?'

'Yes, it's me. Just feel courage to let go. You'll join me in heaven . . . just let go . . .'

The man thought for a minute, and then he said: 'Is there anybody else up there?'

At the time, I was wearing tights with my face on the bum, and banana boots. That might have been my first mistake as far as he was concerned. But generally, people shouted out all kinds of insults. I just had to learn to deal with that – and eventually I used it in my favour. Hecklers could be a hindrance to my show. I learned you have to just nail them right away with a line like 'Don't you get out much?' or 'Did your mother never tell you not to drink on an empty head?', which always floored them. 'Pick a window . . . you're leaving!' was another good one. One night, I was about to play a song when the audience started talking, so I said, 'I want a wee bit of respect, this one's about my granny.' Some drunk guy shouted something about her and went on and on, so I finally said, 'Don't slag off my granny – I won't beat the shite out of you, she will. She's got a black belt in kama sutra, she'll come around and shag you to death.'

Most of the time though, I couldn't hear what they were shouting, so it was just irritating. The audience could understand, but to me it just sounded like loud word salad. Occasionally someone would heckle with a great one-liner, which I never minded. Like the guy who shouted during a concert I was doing in Peterhead Prison in Scotland, which was a facility for long-term prisoners. I was playing the banjo with a great band I took along called Highspeed Grass. As I was playing, *plinky plonky*, my banjo went out of tune. I stopped to tune it and I said to the audience: 'Give me a moment, will you? I've gotta tune this.' A guy at the back said: 'Take your time. I've got nineteen years!'

One of the best things that ever happened to me onstage was in Birmingham Town Hall, which is a great Gothic building in the middle of the city. The audience was going crazy that night, and so was I. My bouncers used to take care of the hecklers; they would throw them out when they got too much. The balcony came right round the building to the stage, so there were people very close to me. There was a heckler about halfway along the balcony being a real nuisance. I said: 'Set the dogs on him!' So the bouncers ran up towards the front seat where he was sitting but he ran and hid from them and they had all kinds of shenanigans trying to catch him. Meanwhile the audience was getting wilder and wilder. The bouncers got him and were dragging him up the stairs. 'Fucking leave me alone!' People were screaming with laughter. The bouncers and the heckler crashed through the door at the top of the stairs and down into the no-man's-land from where he would be flung out, and at that point I went up to the microphone and said: 'His mother's very ill.' It just collapsed the audience, and it collapsed me. I thought it was the funniest thing I'd ever said. It was just a great night. Encompassed everything I ever wanted to be – a wild, funny Rambling Man on the road.

Sometimes it's the venue itself that makes me nervous. At Caird Hall in Dundee, they brought a visitors' book for me to sign. I haven't seen it in a long time but it's the best one I've ever seen. Everybody's in it: The Beatles, David Bowie, the Grateful Dead, right back to the fifties – Frank Sinatra, Tony Bennett. The auditorium is a long narrow hall, so Bob Hope signed, *'The first time I've ever played a tunnel.'* When I read that I thought, 'Well maybe I'm doing something very big here. Maybe this is more important than I imagined.' And then I started to get a bit nervous.

There are some very beautiful halls dotted around the world. But in some of these beautiful halls you find it hard to hear the audience's laughter. When that happens, you don't think you're performing well. There are a few outstanding examples of those 'dead' halls – even really famous ones; I won't say which. Good sound is crucial to a comedian. I use a radio mic onstage. Once I was playing in an arts complex next door to where *West Side Story* was being performed. The performers in that show used radio mics as well, so I was hoping there would be a bit of a crossover. It would have been very funny if the guy singing 'Maria' had bad language interference in the middle of it. 'Mariaaa— Fuck off!'

The Sydney Opera House is the most spectacular concert house in the world. Performing there is like playing the Taj Mahal. In some of the great venues in the world – Albert Hall, Carnegie Hall – once you're in the door you only see the inside. But this place – you can see for miles. I love it. It's right on the harbour. From your dressing room you can see the Sydney Harbour Bridge and Luna Park. And on your approach to it there's a walkway with famous people's names inscribed on manhole covers – Sydney's answer to Hollywood Boulevard, only with Australian luminaries: the poet Dorothea Mackellar, Thomas Keneally who wrote *Schindler's Ark*, feminist author Germaine Greer, and author and broadcaster Clive James.

I like playing to New Zealand audiences. I think New Zealanders are very friendly in general. When I first went there, they reminded me of a very old-fashioned kind of British people with attitudes that didn't exist much any more. But there was something very soothing about that. They would be in their sensible shoes and 'serviceable' beige raincoats, while I would show up in stripy suits and velvet brothel-creepers, and it was perfectly clear which one of us was in charge and which one was the odd, windswept and interesting showbusiness personality. But that's changed radically since I started going there. It's wonderfully diverse and exciting now. Yeah, I didn't know what to make of the place when I first went. It wasn't sunny. It was kind of windy and slightly rainy. I was trying not to feel disappointed. But it did feel rather like Scotland. A lot of New Zealand makes me feel at home. A bad hair day every single day.

———

One of my favourite gigs in New Zealand was on a tiny triangular island that lies thirty kilometres off the South Island of New Zealand. It's called Stewart Island, which where I come from sounds like it would probably be a person. I had the best oysters there. The Stewart Island Town Hall holds 255 people, which is half the population of the whole island. The hall was sold out, so a lot of people sat outside. It was one of the smallest audiences I've played in recent years, compared to Christchurch, for example, where I played to 14,000 people over two nights. I've already explained to you that a huge audience is a great luxury because they laugh longer – the ripple of laughter moves around and there's a space between ending one thing and saying another that is very helpful. It's a bit like surfing. When the applause starts to go down from its highest pitch it's like a wave you need to catch. Just like surfing in the ocean, there's a moment during the wave's descent when you should put your board onto it – or start talking again. But it was still lovely to play the little Stewart Island gig. In many ways I could have been playing Alabama or Edinburgh – size means very little when the lights go out. I know what you're thinking: 'Billy, there's a joke in there . . .' Well, I'm not going there, and you're a dirty bugger.

As was the case in Australia and Canada, my first concerts in New Zealand were largely attended by expatriate British people. Expat audiences can be really wild in some places, and New Zealand was no exception. They were singing Scottish songs before I came on and pretty hard to control, although they were very well worth it when I did manage to take charge. Crowds were sometimes too excited before I came on, a bit like the anticipation before a football match. In a way I was quite proud of that, but it took quite a few years of Australian and New Zealand tours to calm the audiences down. I think the original guys died off and were replaced by their children and their friends, and regular fans. It got better and better, and it stayed better.

I usually ended my show with: 'Ladies and gentlemen of Adelaide . . . it's been a pleasure talking to you and I hope we meet again. Thanks very much.' I'd walk offstage, and then it would always take me ages to wind down. As the end of a tour approached, I'd have mixed feelings about it. I always thought I was going to enjoy the last night but for some reason I never did. Instead of being grateful it was over I'd be worrying about my next tour in the distant future. It's the most misunderstood thing about comedians; after you see him, he has to go somewhere else and do it again . . . and again . . . and again. People who think they are funny because they did two funny lines at their daughter's wedding don't know what it's like when the spotlight is on you and 3,000 bodies are all staring at you, and you have to be funny for three hours for the fifth time that week. That's when comedy really kicks into life, but it takes a long time to get to the point where you can do it to order, to where you can do it with the flu coming on, when the weather's lousy, when you've just had rotten news from home. To be able to be funny under any circumstances – that's the biggest challenge in comedy.

———

I've been lucky to have been helped by so many people on the road. I couldn't have done it by myself. See, I've never grown up. I've never become a man. I'm immature. I'm a manchild. Physically I'm as adult as you get. I'm old. But mentally I've remained a child. I've been looked after all my life – by my sister when I was a child and by my wife and others as an adult. When I do that thing onstage about my sister looking after me and I'm crying all the time – it was based on reality. Children intuitively know when something's right or wrong, and I have some of that childish intuition and it suits me well. I know instantly if I want to spend any time with someone or not, and who will look after me well. I've been well taken care of by people like Ian Magan, Kevin Ritchie, Harley Medcalf, and my long-term manager Steve Brown. So, hitting the road and going on tour again was like going to see my friends.

I'm not ambitious. Any time I see ambition in myself I revolt against it, maybe because any time I've tried to drive myself forward it hasn't been successful, and it didn't bring me the happiness I thought it would. It goes back to the same fear of not fitting in or being unable to keep up with the other boys that I had when I was young. And ambition and success seem a bit contrary to my Rambling Man sensibility; it's as if trying to succeed involves a betrayal of who I am at my core. So, instead of reaching for success, I've always surrounded myself with people who are ambitious for me. It's always their idea that I do certain work. And I don't do things just for the money. I'm not driven by possessions. The philosophy of Buddhism seems to work very well for me. I'm not status-driven either. I think if you do what you do well, certain things will automatically happen for you. I always say to my children: 'You be you, and I'll be me.' I brought four women and a man into the world . . . there should be some kind of prize for that.

Yes, I've been supported by great people during all the years of my touring. I mainly toured with Jamie my roadie, and Malcolm my sound man. We hired people in each town to operate the spotlights, as well as 'humpers' who erected my backdrop. Jamie was in charge of this crew, and he took care of my instruments. He also woke me up in the morning, packed for me, fed me, sorted out my problems, and generally acted like a rock 'n' roll version of Mary Poppins. He used to order stuff for my room and bring it up himself instead of the room service guys. He hated that. He would complain to everyone – 'It's fucking ridiculous. I'm doing a waiter's job.' He was right. For a grown man like me that was ridiculous, but at the time I just didn't want the waiters to see inside my room. I didn't want them to know anything about me. I'm still like that. I don't like answering the phone. Don't know why. I never want to speak directly with people about stuff. Pamela says many famous people hide like that to try to protect their private selves. I just know I really needed that kind of care Jamie gave me because touring was so stressful and took so much out of me, both physically and mentally. I'm a very weird person. Not easy to be with. For example, it was very dangerous for anyone to wake me up in the morning. As I was becoming awake, I would flail my arms and fight and shout – Jamie had to prod me with a broom to avoid being attacked.

————

Somewhere in America, I visited a marijuana farm. No surprise I can't remember the name of the town or where it is exactly. I've never been able to smoke marijuana while upright. The stuff just poleaxes me – so, when I tried it in the seventies, I had to do it lying down. To tell you the truth, I never really liked it. But anyway, I visited this farm where medical marijuana is grown for people in the USA who are legally allowed to use it. A wonderful guy ran the place. He had a comfy bed in the middle of the field where all the plants were growing so he could spend the night breathing it in. He gave me a bottle of cannabis oil and said it would be good for my Parkinson's. I tried one drop, but it was way too much so I gave the rest of it to my crew, who ran off cheering.

That was a wee trick of mine. After the show, I'd spend some time with the crew and then I'd leave early so they could get pissed. It takes a lot of living to remember that you're the boss. I never wanted to be the boss, but I realised during my early concert tours that, being the 'big shot', I hampered their fun. So, on those first tours I did of Britain and Ireland I would be with Malcolm my sound guy and all the crew for a short while, to have a pint or two, then, because their work ethic was 'not in front of the boss', I'd say: 'I'm knackered,' and go upstairs so they could get shit-faced.

———

Malcolm Kingsnorth was my sound man for fourteen years. He had been with the band Status Quo for around twelve years before he joined me. I have noticed that people who tour with rock bands have either been with the group for twelve years or a fortnight. People know right away if they're going to fit in or not. Malcolm was a lovely man. He started as a rock 'n' roll guitarist when he was a teenager. He got to know the equipment so well he became the sound man for his band, and eventually that's what he ended up doing. He was such a good sound man that I never had to bother with soundchecks. I did one at the beginning of the tour and that was it.

Malcolm was fun to travel with. Touring can bring out the best and the worst in people. Once, when we toured Australia, Malcolm's wife had put him on a diet, and my promoter Kevin's wife had put him on a diet too, so they hung out together eating whatever the fuck they fancied. Bad boys. They used to do a 'toast race' in the hotel at breakfast. The bread was toasted on a conveyer belt, and there would be a competition to see whose would come out first. There would be huge roars from the toast machine as somebody won. There was money or other kinds of prizes on it – sometimes pies. Both Malcolm and Kevin loved pies, but they weren't exactly on the diet list. But the shenanigans of our wee group were very tame compared to the behaviour of some legendary bands. I used to really envy those really legendary bad boys. They would usually have a green room with a bar for meeting guests and groupies, and for getting pissed after the gig. I heard about one band that used to show porn movies too, and one night they were fed up, so they decided to project them on the building across the road. Cars were crashing – *beep beep beep!*

———

My promoter Kevin Ritchie was amazing. He was one of the nicest men I ever met. Full of little quirks and twirls. He was always well dressed in blazers and slacks that were perfectly pressed. He was a mad ironer. Loved ironing. He would iron everything he could get hold of. If he had nothing else to iron, he would take down the curtains in his hotel room and iron them. When he ironed shirts, you would think they were fit for a shop window. He had been a fat man when he was younger, so he took care of what he ate. He would eat pies and break the rules, but he never went too far, never risked getting fat again. He was a wonderful person to spend part of your life with. Ritchie is a Scottish name. Maybe he was of Scottish descent. Like me, he enjoyed Indian food.

Kevin and his wife Betty were delightful together. They were like lovebirds. They would tell me little stories about things they did when they were younger. They used to go on weird cruises to places like Papua New Guinea – they'd sail on merchant ships, before proper cruises became popular. Kevin had worked for record companies as a rep selling records and he knew the registry number of every single record the company sold in the fifties and sixties. Kevin was way more than my promoter to me – he really supported my creative efforts. He would notice new things I did onstage. He'd remember them and repeat them to me. He always latched on to some small thing in my stories, some peripheral character or thing that just popped out of my head. A dog appeared once. I said, 'Nobody ever writes about the wee brown dogs. Nobody talks about them or sings songs about them. Where are they going?' Then, when I came off, he would just say, 'That bloody dog!' and start laughing again. He said, 'I never get beyond that in your show.' He was knocked out. He loved my 'wee brown dogs'. And I tried doing a spoof of an opera once. This was at the Sydney Opera House. Kevin saw that piece being born and he loved it. I used to say, 'There's only three songs in the world of opera, and they sing them again and again and again – at different times.' I'd be developing a story and people would suddenly appear – people that didn't even belong in the story, they just showed up out of some nether region of my mind. I would come offstage and he would still be laughing. Maybe he was the only person standing on the side of the stage in my whole life who followed what I did very closely and was really focused on it. That was important to me, in a friendly way. I wasn't up there on my own.

The only man I ever knew who messed with biker gangs was my New Zealand promoter Ian Magan. He was a wonderful guy. He had been a disc jockey on New Zealand's offshore pirate radio. He had two kids, and a wife who called him (high-pitched New Zealand accent) 'Macgannnn?' He was a big man, with his upper half much larger than his lower, and huge hands. He was the son of dairy farmers and when he was a boy he rode a horse to school barefooted. I got to know that rural New Zealand life through Ian and I've never stopped liking it. At one of my outdoor gigs – I think it might have been in Hamilton – a group of bikers were at the back making a bit of a nuisance of themselves. There was a big pond between the audience and me, and a hill sloping up to the back, and they were sitting on a high fence beside a drop. Ian Magan went up and, without warning, lifted their legs and tipped them into the valley. He was no slouch at surprising people and was afraid of nothing. In a former life, I think he may have been Glaswegian.

I remember Ian punching a guy at one of my concerts. A punk guy was constantly kicking the exit door for no apparent reason. Ian intervened, timing it perfectly. He waited until the punk got up to kick the door again and, right on cue, Ian opened the door, ran in and belted him. The punk flew out and Ian shut the door.

Another time, Ian punched Chuck Berry. Yeah, chased him into the airport for apparently stealing money. Chuck Berry got through the customs, but Ian leaned over the partition and belted him in the ear from behind. He was amazing, Ian. I loved him.

My British promoter Mel Bush was a wonderful man too. And so was my late long-term manager Steve Brown. I met Steve when he came to one of my gigs somewhere near Hastings where he had a farm. Steve had retired from running a record company and had become a farmer, but after seeing my concert he changed his mind and began to manage me. Steve was a kind of intellectual hippy type. He liked music very much and had a great mind for the business. He knew about records and how they were made and sold. He knew about rock 'n' roll. He had run the record company very well, and he managed me very well. He was an honest man and I miss him.

———

You really need to have good people with you on the road because you never know what might happen. Once I was filming in Swansea. The place used to be run-down, but it's been rebuilt. They've turned the docks into a very thriving place with cafés and restaurants and trendy shops, and now it's a nice place to walk around. I had wanted to visit the house where Dylan Thomas lived but I never got there. I was always waylaid by something. For example, my Swansea hotel went on fire. The sauna burst into flames. I was in my room, watching TV and eating biscuits. Drinking cups of tea as is my wont. I started smelling smoke, but I just kept watching my programme. Then I heard the fire alarm. I usually ignore alarms, so I still didn't move. Then somebody banged on my door *bom bom bom!* and shouted: 'You have to leave, there's a fire!' I rarely act quickly, and am seldom shaken, but on this occasion I calmly picked up my banjo and my toothbrush and exited my room. I climbed down the stairs, forgetting to take my room key, and joined the throng of similarly annoyed hotel guests who were also stuck on the pavement outside their rooms.

There have been lots of fire alarms going off in hotels where I'm on the twelfth floor or something. Such an irritant. I never fancied climbing down all those stairs, so I rarely bothered. I'd leave it for some time, but the alarm would keep going and drive me mad *Goooga Goooga Goooga!* And finally, I'd just have to get out of there. Sometimes – as happened in Swansea – there was a real fire. Pamela always gave me shit about my attitude. 'Billy – wouldn't it be better to leave sooner rather than later?'

'Maybe. I never bother.'

'So let me get this straight. You smelled burning *before* you heard the fire alarm, but you still ignored the siren?'

'Yes.'

'Okay. Not such a great plan, Billy.'

'Well, it was only on the first floor and there's always a window . . .'

'BILLYYYYY!!!'

I had a fire emergency one night while I was playing Hull. We had cannons onstage and at the end they scattered confetti all over the audience. *Boom boom boom!* But one night they went off early by mistake, in the middle of the show. My manager came onstage to tell the audience everything was okay, but during his explanation lots of guys in yellow coats came running in with axes and fire hoses. I said: 'I'm sorry about this.' They were complaining. They said they'd been watching that John Wayne movie *She Wore a Yellow Ribbon* and had to miss the ending.

———

All sorts of unexpected things happened while I was performing. Sometimes people just came wandering onstage. They'd go to the toilet and get lost, then come in the wrong door and suddenly find themselves onstage right beside me. They were always amazed to see me, but I enjoyed the interruption. I usually just talked to them. 'Hello! How are you doing? You've come in the wrong door. Say "Hello!" to the people!' Then they'd scurry off. Sometimes dogs would come in. That happened a lot when I used to play outdoor festivals. Once I was onstage playing with the Humblebums in a rock festival in Halifax and a dog wandered in and peed on my leg. The audience collapsed. They saw it lifting its leg against my velvet flares before I knew what was happening. Mark Knopfler, the Dire Straits musician, was there. At the time he was reporting for a Newcastle newspaper – he wasn't yet in a band. Years later, when I met him in Australia, he mentioned it.

Size Means Very Little When The Lights Go Out

The more informal your venue, the more likely it is that something unexpected will happen while you are onstage. For example, playing outdoor festivals can be weird, especially at night, because the audience gets into sleeping bags. They might be applauding heavily, but the sound is muffled inside their sleeping bags and you don't hear it. You may think you're dying the death while you're actually doing rather well. And there are other festival hazards. I remember one night the band the Pretty Things – an ugly, hairy group – was playing on the same bill as Gerry and me, and somebody got electrocuted. The tarpaulin over the stage became waterlogged with the rain and burst. Water came skooshing down and hit the electric wires – *zap!*

Like I say, I'm no stranger to accidents. Personally, I think the best cure for any kind of accident or injury is aromatherapy. Smell yourself better. Imagine, there's a huge accident on the freeway. There'll be a hundred trucks piled up and blazing away, people lying with bits missing, 200 cars in the same condition and a pedestrian who caused the whole fucking thing running out on the road after his cat. There are flames, sirens, policemen, paramedics, black smoke filling the sky. Through the crowd comes a man:

'Let me through! I'm an aromatherapist.'

There's a truck driver lying against the wheel. His leg's missing.

'Listen, I'm just going to rub some of this lavender on your stump . . .'

'Agghhh!!!'

'That's right. You're relaxing now. See – you're leaning over to one side . . .'

'I'm NOT relaxing! I'm reaching for my leg. I'm gonna fucking hit you with it!'

9

PLAYING ELEPHANT POLO IN INDIA

—

ONE THING AN actively travelling Rambling Man cannot do easily is keep a pet, although you can get stuck with dogs sometimes. You don't find them – the dog finds you. Follows you around for a while then eventually disappears. I can't imagine trying to jump on a train with one of my lazy little dogs I have at home. At the crucial moment, they'd sit down in protest like they do when I walk them. 'I'm not going a step further, Billy. You'll have to carry me home.' If I was ever getting mugged I'm sure they'd just roll on their backs hoping for a tummy rub. Yeah, unless you have a caravan you can't really take a dog on your travels without courting problems. But some guys I knew had rats. They kept them up their sleeve and you'd just see the tail sticking out. They loved those rats, and they were beautiful. Clean as a whistle.

Even though they don't usually travel with them, Rambling Men encounter animals every day, either through picking up work on a farm, passing by stables or travelling through places with all sorts of creatures. Rambling Men are typically interested in wildlife, it speaks to their curious nature. One of the best towns I ever visited was a place called Rayne in Louisiana, which was overrun with huge frogs. I never saw frogs that size before.

Instead of just trying to kill or eat them the people there held all kinds of events based on it being the frog capital of the USA. They even crowned a Frog King and Queen every year. It was great the way the people liked the frogs, with wee girls dressing them up for competitions. And it was nice to capture that on film and show people *that* hidden corner of America – the America where they don't shoot people or join the Ku Klux Klan, they just get their frog ready for the morning. Delightful.

———

In Texas, cows outnumber people two to one. I went on a cattle drive when I was filming there, with real modern cowboys. Baseball hats and jeans. Stan the owner wore a ten-gallon hat. No wonder cowboys wore bandanas, being stuck among all those farting cows. I heard that cows do more damage to the atmosphere than cars. But, despite the terrible smell, I did feel like John Wayne riding on my horse, herding the strays. I imagined doing that a million times in my head when I was a boy, smacking my own backside as I galloped along the Hyndland Road. Cowboy movies sparked my original fantasies about escaping my life in Scotland, to roam the prairies on a trusty steed and camp in the canyons with my posse of fellow renegades.

When I met real cowboys, they didn't exactly fit my boyhood fantasy at all. In Texas I found them to be angry people – politically active. They didn't like oil wells being dug on their land – and they didn't like the people who protested about the oil. They were on nobody's side. They didn't like lefties poncing about in the countryside telling them what to do, and they didn't like big business drilling holes without their permission. In my memory, they previously used to be a stable, pleasant crowd, but if they've changed it's understandable because they've been messed about for years. They would take the piss at me playing my banjo or guitar. 'So, you're a cowboy, are you?' 'No, I just like the music.' 'Yeah. You're all the same, you guys . . .' There's an edge. Well, we were outsiders coming in and playing music in their town hall – music that was part of their culture, not ours. We were singing about their way of life, but we weren't part of it, weren't one of them. They've got a right to feel like that. It didn't seem to count that we were there because we loved the music and admired them; to them we were just fashion followers, glitzy, showbiz rhinestone cowboys.

I wrote my own comedy version of 'Rhinestone Cowboy' once; it went like this:

I'm a half-stoned cowboy,
Who fell off on his arse in the middle of the rodeo,
Just a half-stoned cowboy,
Buying beer and whiskey for people I don't even know,
And I wonder where my money goes.
I was having a drink alone,
When someone called me to the phone,
He was making a joke about trouble that was coming my
*   way,*
It was my wife saying what's your game,
You know you ought to be bloody ashamed the way you
*   carry on,*
You got a load of children starving,
And a jungle that once was a garden,
And you're off like a shot every night as far as I can see.

————

(Everybody! . . .) *Like a half-stoned cowboy . . .*

———

Maybe the cowboys I met had done their research. Maybe they were wary of me because I'd joked about them in the past. I used to make a noise like a train whistle onstage and say: 'That's a lonesome whistle. Cowboys like them when they're sitting in the prairie at night, guarding the steers. Singing songs over a blazing Apache. Eating beans and farting all over the place. They hear a lonesome whistle in the distance, and it reminds them of their true love far away. They shed a silent tear, and then they get into bed and interfere with themselves. Cowboys are very short-sighted people.'

———

The first time I rode a horse for real was for a TV show where famous people were taught to do new things they hadn't been capable of before. One guy learned to hang-glide, and I learned to ride a horse. I went to some posh stables in Kingston-upon-Thames. On the first day, the horse stood on my foot. Fucking painful – I'll never forget it. I had no idea how near or far you could stand from a horse. She was a lovely horse, a tremendous-looking gold-coloured mare, but I think she could spot a rookie. She bit me one day. I was standing outside her door – you know those half doors? And she leaned over and apropos of nothing just bit my shoulder. I went, 'Get out of it!' very sternly and she never came near me again in a malicious way.

The first time I sat on the horse it was indoors in the riding school while my trainer led us round in a circle. Eventually I graduated to riding round the park and crossing the pond. The water was up to her knees. Then we did a few jumps. That was hard because I couldn't control my bum properly. When you're jumping and airborne, you're supposed to stick your bum out backwards. At first it felt really awkward, but eventually I got the hang of it.

I slapped the horse's neck one day. The trainer said: 'We don't slap horses' necks in here. That's what cowboys do in cowboy films. You can pat her if you want.'

What I remember most was just walking through the woods, and there'd be squirrels on a tree just looking at you – they didn't run away cos you were a horse, you weren't a guy. It was great.

When I'd learned to trot and canter through the park, I was considered ready to be on film, so a crew came to capture me riding through the New Forest National Park near Southampton. It's a place where they train wild forest ponies. Some of them become working ponies, while others just stay in the forest eating grass and fucking about. It's very different from the Kingston stables. The people who look after them are called agisters. I was stood talking to my horse one day and she bit me. Second horse, second fucking nip. One of the agisters came over and said, 'Don't let her do that!' and punched the horse right in the face. I thought, 'Oh, this isn't Kingston.' The horse went, 'Oh fuck.' Took the lesson seriously.

The agister said, 'Get on.' I mounted the horse. He said: 'What have you done?' I said, 'I've trotted and cantered, jumped a bit. Not much.' He said, 'Come on, we'll see.' And he slapped the horse on the neck from his horse and we shot up a hill together. It was exhilarating. Through the trees to the top of the hill and along the top. He went *Swishhhh!* and hit it again. That's when my horse went cylinder-shaped. I crossed my legs underneath her. I felt like I was on a wild rocket ship. We rode along to where the other horses started to appear, and by the time we got close to the end of the park there were horses everywhere – about forty of them. They started running along the fence beside us: *'Ayeeeghhh! Ayeeeghhh!'* Neighing loudly and making quite a racket. I was truly living my rambling cowboy fantasy now. We were going like a fucking train. Then we cut down a hill, and as we picked up speed I could see a branch coming towards us about the horse's height. They had told me to watch out for that at the riding school so I just laid back on the horse's bum and went under the branch, and the people from Kingston went *'Whoohooo!!'* I felt like a ruffty-tuffty stuntman.

———

The next time I rode a horse was for a movie starring Tom Cruise called *The Last Samurai*. I played Sergeant Zebulon Gant, an Irish American Civil War veteran. They said to me 'Can you ride a horse?' I said, 'Yes.' They said, 'Can you do military stuff?' and I said 'Yes', although I didn't have a fucking clue what they were talking about. It was easy. I just had to keep my horse walking in line with the other guys. The horses liked doing that. We had to keep them steady and make sure they didn't panic while we rode beside a burning building. I had to ride down the hill with several of the men to the 'headquarters', looking as if I'd done it all my life, while holding fire weapons. In the battle scenes, I had to scream orders: 'Stand your ground!' 'Return fire!' to the men while samurai were attacking them through the trees. That was terrifying. Fortunately, my horse liked me.

We filmed all this in New Zealand, and I got to go fishing in the middle of it. Yippee! You know the loveliest thing I did in New Zealand? I released a wee kiwi into the wild. It ran happily into a protected area where none of its predators hang out. Delightful. Kiwi Whisperer, that's me. It was such a joy to watch it scoot away. I've had wonderful encounters with all sorts of creatures – koalas, wombats, penguins. I'm the Joy Adamson of comedy.

———

In Malta once I was walking along a cobbled street and I heard a fast *clippity clop* behind me. A fast *clippity clop*. I turned round and there was a guy exercising his racehorse by holding its reins out the back of a moving station wagon. I was peeing myself laughing. I'd never seen anything like that before. They practise controlling the speed – twenty miles an hour, twenty-five miles an hour, whatever, coming down long straight roads. He took umbrage because I was laughing. 'Fucking Scottish prick!'

Controlling animals is often a lost cause, as I was firmly reminded at the Payson Rodeo in Texas. It was heart-stopping. The first guy I talked to was a veteran rodeo rider. 'I used to ride bulls for a living,' he said. 'But then, three broken necks later . . .' He told me that bulls are incredibly smart animals. 'They learn your moves the first time you ride them, and then they're way ahead of you. Know all your tricks. Seventeen hundred pounds of killer intelligence – it's like climbing in a ring with Muhammad Ali.' It's a hard life. The rodeo riders travel hundreds of miles, and then they have to pay to compete. And if they don't win, they don't get paid. I met a woman who said she was the 'Rodeo Queen' and the World Champion Barrel Racer. I had no idea what that meant. She explained that the competing riders raced their horses in the arena, manoeuvring around three barrels that had been placed in a triangle formation. They were astonishingly skilful at it. Went like the clappers; they say you have to 'ride it like you stole it!'

All ages were involved in the rodeo performances. I saw a five-year-old performing crazy rodeo moves on a hyperactive sheep – and getting tossed off. Brutal. In 1968 a ten-year-old known as 'Annie the Okie' actually won a barrel racing competition but didn't live long enough to receive the gold buckle. She was killed in a highway traffic accident while travelling to a rodeo event between Little Rock, Arkansas and Waco, Texas.

———

When you try new skills while filming TV programmes you can really get shown up. I practised archery to go hunting but only managed to hit a plastic moose in the arse. The next day my hunting guides, twin sisters Sherry and Caroline, showed up at 5 a.m. to take me turkey hunting. I'd never hunted wild turkey before, but at least in my full-body camouflage I was dressed to kill. When we got to the hide, I disgraced myself again by getting the giggles when one of the twins made a very realistic turkey call to lure their prey. They worked together: Sherry tried to lure them with a call, then Caroline attempted to shoot them with her bow. That was the theory, anyway. It was uncomfortable there in the hide. My bed had been so warm. Dawn broke . . . and still nothing. Turkey sandwiches were no longer on the menu. Frankly, I was delighted they got away.

I saw American buffalo during my train tour. I love them. They've got enormous heads. And when you walk up to them, they're covered in flies. You wave your arms, and they fly away, and you think you're doing the buffalo a good turn, but it kind of likes having the flies around. When you see them in Africa, they're covered in wee birds that eat their parasites. Buffalo are beautiful creatures, but they can be a bit uppity. When you're fishing, they get nosy, and they come along to check you out. They've been known to flip people over their head. I was fishing with my son Jamie in Yellowstone National Park and the buffalo came down. People told us, 'Watch them, they're a bit . . . territorial.' Glorious-looking creatures.

I would have liked to see more wildlife when I travelled in Africa, but I never went on safari. Quite the opposite. The first time I was in Biafra for work – I was a welder then – the only animals I saw were men who'd been on a dry oil rig all week, at a bar on a Friday night. I include myself. The other times I was in Africa were to film for Comic Relief and there were far more pressing things to focus on than enjoying a trip out to the bush. Talking about wildebeest and lions in my concerts just came from watching wildlife programmes on television. I loved them. David Attenborough is my hero. I was on the *Parkinson* show with him a couple of times. He once walked on to the show, sat down to discuss a new programme he'd done on birds and turned to me and said, 'Do you know about Penduline tits?' I gave him a sideways look: 'Are you offering me something here?' and he just burst out laughing. Another time, he told a story about going to the toilet in some foreign country and a rat came up between his legs. I said, 'I hate it when that happens,' and he just lost it. After the show I was singing a song and we danced up the stairs together. What a life he has had!

I did meet some African wild dogs in a wolf sanctuary in Missouri and fed them a few pigs' ears and a rat. They saved the rat's tail for last – apparently, it's the best part. Those dogs are amazing creatures. They share their food, and when they go hunting in a pack they leave a babysitter behind to take care of their young. When they return, they regurgitate to feed not only their babies but also the babysitter.

The sanctuary existed because the wolves were nearly extinct around there at the time. There were forty wolves, and the keepers were careful not to pet them, talk to them or play with them because they were committed to keeping them wild and afraid of humans for their own protection. I went into the wolves' enclosure and helped feed a deer to them. One wolf had interesting table manners. He peed on the food first – maybe that was his way of defrosting it.

———

In the USA, quite a few African animals end up in private hands; in fact I was told there was a 50-billion-dollar illegal industry involving at least fifteen thousand big cats. I think Americans only learned that after the widespread popularity of the TV series *Tiger King*. I once met a baby tiger who had been rescued. His name was Anthony, and I fed him from a bottle. Monkeys, llamas, lions and tigers are also kept illegally, but they are all dangerous pets. I heard about one monkey who had bitten the owner's girlfriend's ear off. I'm totally content with my three wee doggies.

Some wild animals gravitate towards humans just because they can. In the shallow sea and canals where I live in the Florida Keys, I often see manatee. They bask and play around, eat the seagrass and try to find freshwater run-off to drink. I love them. They're massively friendly, although you're not supposed to have any interaction with them. Sometimes they come past my dock with babies and look at me as though to say, 'Meet my new wean!' And in affluent areas of America where people have heated pools, hot tubs and jacuzzis, people are always posting stories and videos about bears climbing in and relaxing in the warm water, especially in the Great Smoky Mountains. One man in Altadena, California reported he saw a bear in his hot tub. At first it was just relaxing, but then it started playing with his chlorinator, and tossing the thermometer in the air. Then it loped over to a half-finished margarita on a garden table, knocked it over and lapped it up.

When I travel around the Florida Keys, I notice a lot of interesting animals that most Glaswegians never see live. Many local Floridians are annoyed by the iguanas because they shit all over your yard, eat your flowers and can be dangerous for pets. But I love them. They have the most beautiful colouring – especially the large, bright-orange males. Unfortunately, they get disoriented when tropical storms or hurricanes knock down trees. In their society, the alpha iguanas get to live on the highest branches, but when they get broken off their world is turned upside down. After a big storm you see them wandering around aimlessly, having lost their status as top-branch dwellers.

There are chickens everywhere in the Florida Keys. They are a protected species, so you're not allowed to catch or eat them. The roosters are particularly beautiful, with wonderful, iridescent feathers that catch the sunlight. People have to stop their cars for chicken families all the time. Why do the Florida Keys chickens cross the road? To annoy drivers. And you see road signs in south Florida that are just pictures of crocodiles or alligators. You get no more information than that. But it would be useful to know if those reptiles like to take a wee stroll thereabouts – or are they hunting for lost motorists?

I get that we are asking for trouble when we enter an animal's habitat, but many animals seem to think they can come into ours no problem. I was in an AIDS ward in Mozambique, Africa, during the filming for Comic Relief. I was visiting a guy whom we had filmed a couple of days earlier and I went back to see how he was getting on. While I was talking to him a goat came in and pissed against his bedside table, then walked out again. And I saw a giraffe at Nairobi Airport. The park is adjacent to the airport. I said to the taxi driver: 'Look! A giraffe!!' He just gave a disinterested 'Yeah?' like I'd pointed out a pigeon.

I never liked pigs. I find them rather vicious. I was bitten by a pig once, when I was just trying to be nice to it . . . although I was in its pen. I have become immensely fond of some animals, though. I love my wee dogs at home. And I adored Ralph McTell's parrot. His name was Albert, and he was the best talking parrot I've ever come across. 'Hello, Bill!' If you rattled your keys he'd say: 'Is that you away, then?' And if the atmosphere was quiet, he'd say: 'What d'you have to do to get a drink round here?' Ralph became really worried when Albert developed a cough. It was a deathly rattle, so Ralph took him to the vet. The vet examined Albert, then said: 'I hate to tell you, Ralph, but that's YOUR cough! He's parroting you. Your parrot is perfectly healthy, but you're going to have to do something about your own cough . . .'

———

Australian animals frighten the life out of me. They've got a larger selection of creepy-crawlies than exists anywhere else in the world. I've suffered a couple of spider bites. I had a sore on the inside of my leg and didn't know where it came from. Turned out that, unbeknownst to me, some vicious little bastard had sunk its fangs into my leg. I had to have it drained and looked after. This was in Perth, Western Australia. While the doctor was administering it to me, Pamela came in and threw a Mintie at me. I think she thought she was cheering me up, but I was in agony. She said it was the local cure for a spider bite. She grew up with deadly spiders in her backyard, so for her it was like having a splinter removed.

People around me often underestimate my arachnophobia, but not my Australian promoter, Kevin. Once when we were going to a gig I got out of the car and there was a spider on my shoulder. Quick as a flash, Kevin grabbed it. I didn't see it, or I would have fainted. I was bitten another time when I was climbing down some exterior stairs of a house in Pearl Beach, running my hand down the banister, and the wee bastard got me from underneath. Savagely attacked me. I was sure I was going to die. See, Australians live in paradise, but they have a lot of horrible creepy-crawlies. They're everywhere. Any country that has a poisonous spider that lurks in the lavatory just waiting for you to sit down is a dangerous place. Not my idea of a good time, being bitten on the scrotum. Or anywhere in the genolica area. 'Doctor, you'll never believe this . . .' And there was an Australian woman on a nature programme I saw. She bred funnel web spiders. They are the Glaswegian hard men of the spider world. Aggressive and lethal. She actually seemed to like this spider. She said: 'He's nice!' Fuck. Imagine him standing on his hind legs: 'C'mere!' Just imagine his fangs sinking into one of your testicles while you're sleeping.

I love the bigger Australian animals though – kangaroos and koalas. They're beautiful and, apparently, they taste good. Not the koala. I saw kangaroos in the outback as I was travelling to Iron Knob. We were driving along, and kangaroos just started hopping along beside us. I don't know why. It was like dolphins following your boat, or the wild horses that followed our car in Montana. Maybe the kangaroos thought we'd feed them, like the monkeys that used to climb on your car and rip off your windscreen wipers at Windsor Safari Park. Anyway, whatever they wanted I thought it was amazing to see that. It was the first time I saw a live kangaroo.

My favourite animal in the whole world is the wombat. I first saw a wombat on a TV programme. He was a hairy-nosed wombat and I thought that was the nicest name for a creature I'd ever heard. Later, I met some people in Australia who told me they had a wombat as a pet and that he was a downright nuisance. He dug huge holes in the garden, which made them furious because he wouldn't come back through the same hole – he'd always dig a different one. Their backyard was just a mess of underground warrens. But the contrariness of the creature really appealed to me. Wombats have a wee pocket just like kangaroos but it's the other way round – the opening is near their legs – so it doesn't fill up with stuff when they dig. The first live wombat I saw was in a TV studio with his zookeeper and I was allowed to pat him. He was like a big mouse. I never saw one in the wild, but I had a toy stuffed wombat called Wally. I saw a few in the zoo, and I really wanted to get closer to this one they called 'Digger', but they said he was a vicious bugger. Another smaller guy was called 'Not So' because he was not so hairy as the others. Really? They must stay awake all night thinking up these names.

I like platypus too because they are creatures with features that normally shouldn't go together. It seems like a design flaw. A beast that lays eggs, fights with his claws and has poison on his back? I've never seen one in the wild though. Nowadays I like to draw weird creatures; maybe they're inspired by the platypus. I invented the Gozunder Fish that has a unicorn horn and a kind of platypus tail-shaped leg, and I drew another odd fish that has two legs – 'Run to the Fly'.

———

My parents used to accuse me of being 'Up a gum tree'. I never knew what it meant but I gathered it wasn't good. Now I think it's a great thing. I like the idea of being up there with the koalas – stoned, and shagging night and day. Koalas are the hippies of the animal world. There's an island I visited off Perth called Rottnest Island. I flew there with a film crew. It was a bumpy flight and, on the way, I was sick all over my jacket and the microphone. Ever the consummate professional. Anyway, on Rottnest Island you can see these little creatures who live there that are called quokkas. When some eighteenth-century Dutch explorers saw them, they thought they were rats, but they are actually wee marsupials. The way their faces are structured, they look like they are smiling, so tourists try to get selfies with them. They are very friendly, and unafraid of humans, so it's not too difficult to get a quokka selfie – although they can bite. Unfortunately, people sometimes feed them chips, which is very bad for them. Another island I visited off Australia is Fraser Island. They have the purest breed of dingoes that exists in Australia. I saw them at Eastern Beach, where the roads are all made of sand. I sat under a tree watching an amazing sunset and playing my banjo with dingoes all around me. Heaven. My God, sometimes being a Rambling Man on the road is hell.

Animals have no idea what month it is. I was filming outdoors in Australia once and I was irritated by March flies. Nightmare. They were dive-bombing me, biting the shit out of me. I wanted to get one of them and shout at it: 'It's not March, it's fucking November!' They should do an airdrop of fucking calendars. Quite a few flying things have got on my tits while I was filming. I was doing a TV presentation in Cardiff Bay once, and the birds were squawking so loudly we had to stop. All the statues were covered in wire to stop the birds shitting on them. To tell the truth, I think it's a bird's duty to shit all over people and monuments.

———

I travelled by riverboat along the Mississippi during one of my American tours, and I'm glad I did. The river was beautiful. Huge. First I sailed on the delta, then peeled off on a wonderful bayou – a tributary full of jumping fish . . . and snappy monsters. I'm rather fond of crocodiles now. There's a few around where I live in the Florida Keys. My friend Chris is expert at catching and relocating them. They're shy creatures – not at all like the ferocious salties you get in Australia that will chase you down the street soon as look at you. But people in the Keys don't really like it when those docile crocodiles come in your backyard and eat your pets, or set up home in storm drains and refuse to leave. Some even lie across the airport tarmac so planes can't take off, and they're not so great for tourism. Then Chris comes along and lures them with a tasty dead thing, catches them, ties them up humanely and loads them in his truck. He takes them home and plays loud heavy metal music to them all night to remind them not to come near humans, then drives them up to the Everglades or somewhere far away to release them to their new home. Wonderful plan.

———

People in northern Australia attract large reptiles by dangling pigs' heads and eye fillet on a string from a pole. I saw this on the banks of the Adelaide River at the edge of Kakadu National Park. A big crocodile came cruising up. Nick was his name. He really liked his eye fillet. People coaxed him close to the bank and I had a go with the pole. Fuck, he came half out of the water to grab it. Jaws open incredibly wide. Two hundred million years he's been around. In 200 million years you get pretty good at what you do – and in another 200 million years you'd be good at anything. I imagine I'd be able to play the banjo properly in 200 million years.

In Cairns, on the north-east coast of Australia, they installed a swimming pool off the beach that had been designed to float in the sea. The thinking was that it was the only way you could swim in the sea and not be eaten by a saltwater crocodile. They launched it with a lot of fanfare . . . but the following morning the local newspaper featured a picture of a crocodile climbing into it. Brilliant. Of course, my Arctic Circle nemesis – the polar bear – is supposed to be one of the smartest creatures on the planet. They are amazing animals, but do they have wifi? No. So they can't be that clever.

Sea creatures are smart beasties. At an aquarium in Auckland, the people in charge told me they'd had some crayfish in a tank across from this octopus's tank and they were mysteriously disappearing. They tried many ways to find out what the hell was going on and eventually discovered that this canny octopus was sneaking out of his tank at night and octopussing his way into the crayfish tank and eating them. He looked very satisfied the following day. Apparently, an octopus can squeeze through a hole the size of a twenty-pence piece. Don't you love them though? The most extraordinary creatures. An octopus has three hearts. You can break his heart and he'll still be in love with you again – twice! And eight tentacles. I thought it was *testicles* when I was a boy. I thought, 'Them buggers must have a scrotum like a bag of potatoes.' And they have suckers you can't escape from. They can drag you into a cave and shag you senseless.

There's a constant battle between man and sea-beast. For example: the stonefish looks like a stone, lying there in the sand saying, 'Stand on me – I'm just a wee stone.' But if you stand on this thing, they say you'll feel the worst pain known to humans. How they know that I'll never know. Maybe they have a pain-ometer. Some kind of meter that goes from 'Ouch' to 'WTF'? From 'Jesus Christ!' to 'Sweet Mother of Jesus' to 'Agony' to 'Worst Pain Known to Man?' 'Fuck! It's right off the scale!'

During my first visit to New Zealand in the late seventies, I saw an article in the newspaper about something that happened during a family cricket game by the sea. They'd set up three large twigs as a wicket in the sand. At one point, the ball fell in the water and the kids ran in after it, but then someone noticed an approaching shark. The father of one of the kids immediately ran into the water. The waves were over his head, but his arse was just a blur. This man got between the shark and his son and whacked the shark in the face with a cricket bat. When I heard that I said: 'I like it here.' It was the kind of thing they put in British comics when I was a boy – crazy, heroic acts. They should make films about this guy.

I'm glad not all sea creatures are threatening. They have delightful penguins in the Auckland Aquarium. I went into their enclosure and sat beside them, and they were rather friendly. One of them was called Fatboy. He was the boss. He sang me a song – what a racket! And in Western Australia I did something

I'd always wanted to do – I swam with wild dolphins. I was a bit worried that there might be other things hanging around the dolphin, though. Sharks don't come charging into Woolworths and bite you in the arse when you're minding your own business, do they? When you hear on the news: 'A shark ate a person last week', where was the person? – in the fucking sea! Well, tough, you're in their territory. But I got up at 7 a.m. – which is the middle of the night as far as I'm concerned – and I went in the sea anyway. I got an underwater scooter and swam with wild dolphins. It was astonishing. Beyond belief. It was my fifty-third birthday. I got a cake with a picture of me swimming with the dolphins! I couldn't have wished for a better birthday, because as a boy I always fancied myself as Jacques Cousteau. To me he was the original underwater Rambling Man, roaming around the oceans, discovering extraordinary things and surviving on his wits. He invented equipment to allow him to explore parts of the sea no one had previously fathomed – a giant among Rambling Men. Unfortunately, my early attempts at following in his footsteps as a scuba diver revealed that, due to my inability to focus on important stuff like how much air I had left, I was best off remaining topside.

———

Australia's got to be the luckiest country on Earth. When one of their prime ministers was apparently eaten by a shark, people in the UK went: 'Why didn't we think of that? All those years we had to put up with Margaret Thatcher – and all those sharks doing nothing . . .' There's all kinds of advice for fending off sharks. People in Australia say: 'If a shark gets too nosy, punch it on the nose.' And 'If a shark bites you, it's just mistaken identity. It would just take one bite and spit it out.' Fuck that! My arse is gone . . .

Sydney Harbour is full of sharks. I went across the harbour on 'the *Bounty*', which is the replica sailing ship they built for the movie starring Mel Gibson. I played my banjo on board, and it was lovely. Pamela gave me shit for mispronouncing Vaucluse, which was the place where an eccentric Irish baron called Henry Browne Hayes was brought as a convict for abducting a Quaker heiress. He had a consuming fear of snakes, so he brought 300 barrels of peat from Ireland and placed it round his place in Vaucluse, believing that no snake would dare cross a piece of Irish 'Holy Land' (it was believed that St Patrick had banished snakes from Ireland). But a portion of Australia without a single snake? Not too likely. And Australian snakes can sneak up on you. Bastards. Personally, I prefer the American rattlesnake. It may be camouflaged, but it's the only snake in the world that will tell you it's there. Give you a wee warning. *Sssscchhhhlackleacklackle*, which means 'Don't make any sudden moves.'

On the *Bounty* we passed Fort Denison – a sturdy stone island fortress designed to protect the harbour. It was incorrectly built; once it was finished, they discovered that they couldn't fire muskets or guns from it! I love the human race. Such wonderful stupidity! We also passed Goat Island, originally a little penal island. It was famous for a prisoner from Glasgow called Charlie Anderson. He made such a nuisance of himself by constantly trying to escape, they sentenced him to two years chained to a rock. The other prisoners carved him a couch on the rock that's now called Charlie Anderson's Couch. His chain was twenty-six-feet long, and when people sailed by he would harangue them. 'Go fuck yerselves!' A man after my own heart. They were so scared of him they handed him his food on a very long pole.

———

Talking about long poles, I have little interest in polo, but when my manager said, 'How would you like to go to Nepal and play polo on an elephant?' I said, 'I was just about to ask you if I could do that.' He laughed, and then I said, 'What are you fucking talking about?' He said, 'It has come up. Someone has dropped out – would you like to step in?' I said, 'I'd love to go to Nepal.' When I arrived at the famous hotel called Tiger Tops, I found that, although we were supposed to have been supplied with polo gear, there was none for me, and my luggage had been stolen in India, so I had nothing. Everyone else was in smart breeches and high polo boots, but I was stuck with khaki pants and a pair of desert boots.

Next day we went to the polo field. It was in the foothills of the Himalayas, with stunningly beautiful scenery all around. I remember being flabbergasted. I looked at one of the mountains and I said to one of the guys who knew what he was talking about: 'Look at that! It's right beside us!' He said: 'It's many miles away.' By the sheer size of it, I thought it was just a few steps away.

The celebrity team was me, Ringo Starr and his wife Barbara Bach, Steve Strange and Max Boyce – what a weird collection of people! The other team was made up of proper polo players, Nepalese guys who worked for King Mahendra, the king of Nepal. It was he who hosted the game – sponsored by Cartier. I think he just liked to hold celebrity matches for no reason at all. The elephants were his. Understandably, the other team all hated us novices because we knew nothing about polo and even less about handling elephants.

I was told to get on my elephant, but nobody showed me how, so I just improvised. I ran at it from behind, jumped on and put my feet behind its head – I knew you did that last bit. But then the guy in charge of my elephant – the mahut – hit the elephant in the head with a file. I said, 'Hey, no more. Don't fucking hit it with that!' Then one of the guys in charge came over and mumbled something. He was probably saying, 'He's a townie. He knows fuck all about it.' But I did know you don't hit animals in the head with files. Later on, I was on my elephant, and a guy was collecting my elephant's shit in a big sheet, and he said to me, 'It's a normal job here.' I said, 'No, it's a nail-biting cure.' They liked me from then on and I liked them. A bit of humour can always break the ice.

We got beaten in the end. Well, of course we did – none of us knew what we were doing. Nobody really showed us. They just said, 'That's your pole. That's the ball. Use the pole to hit the ball towards the goal. The elephants will get very excited but don't panic. They'll rise up on their hind legs and make an enormous din, but you'll get to love it. It was true. The elephants were bellowing '*Bahoooooo!! Bahooooo!!*' when they were running for the ball. My elephant would rear up on its hind legs like the old circus elephants, but I was well and truly tied on by a rope going over my lap, under my thighs and round its belly. The mahut was in front of me, further up the elephant's neck, and I had to tell him which way to go. He did the steering. At the end, we got big fancy prizes. I got a beautiful ashtray. And that was it.

The day after the match they took us out into the jungle, but we'd not been travelling for long when the guide suddenly stopped. 'I can smell a female tiger. She is near.' I said: 'Well, why the fuck are we still here?' We retreated sharpish. Later that day I saw a baby rhinoceros. It was just magnificent, walking behind its mother through the elephant grass. Then, we got off the elephants and we were having a cup of tea, and I saw a rabbit. I said: 'Oh God, look, a rabbit! At last, something that won't eat you.' But as I walked over to it, it scuttled off and I went: 'Fucking hell! That ain't no rabbit!' Its legs were like my fingers. It was the biggest, scariest spider I'd ever seen.

10

WHERE MEN WEAR SKIRTS AND NO KNICKERS

———

———

I DON'T HAVE much of a head for history or politics. As a matter of fact, I don't have much of a head for anything. People have often said to me, 'You should get your head examined!' Well, I did – and was found wanting. Apparently, I have ADD. Attention Deficit Disorder. Fair enough. I don't care. I don't feel any different from when I thought I was stupid, which was my earlier diagnosis. I left school with a letter from the headmaster saying, *'Billy Connolly was always punctual.'* So, I have A level punctuality.

But like other Rambling Men, I am interested in the world. Ideally, the Rambling Man should be a philosopher, a historian and a political theorist. He has a set idea of how the world should be and he doesn't mind telling others – in the right situation. Naturally, the Rambling Man is far too smart to bring up hot-potato topics in the company of people who could offer him a hot meal. People often try to pin me down on subjects like the monarchy or Scottish independence . . . but I never bite. Not this soldier! The Rambling Man has a healthy amount of cynicism, but he also believes the world should be blissful. People should be nicer to each other and more honest. They should place less trust in those who seek power, and support honest people who are good at what they do, not those who will just say anything to get people to vote for them. In an ideal world, we should be able to spot dishonesty a mile away. We shouldn't be so gullible. And we should pay attention to the important stories of people we meet.

For example, during one trip, I met a group of women who were sent to jail when they were children. On 18 April 1959, 26,000 high-school and college students marched to Washington DC to demand the end of segregation, and in particular the end of segregated schools. It was a very important civil rights march, known as the Youth March for Integrated Schools, and speeches were given by Martin Luther King Jr, Daisy Bates and many others. I met three women who were among the 3,000 children who were put in jail that day. I had not previously heard about this, and I was astounded that it happened in my lifetime! It was very moving to hear their story. The women were so pleased to have marched, and they were surprisingly lacking in hatred for the people who had been prejudiced against them. As little girls, they had endured a great deal of prejudice and segregation – passing the showground but not allowed in – I would not have known about this if I hadn't actually travelled there.

Here's another thing I learned: The state motto of New Mexico is 'It Grows as It Goes', whatever that means, but during World War Two – at least for the town of Los Alamos – that probably should have been the other way round. Nobody at all was allowed to enter the town after 1943 without military clearance. Since then it has been a town of mystery, going quieter and quieter as it grew to be the home of a secret government-funded programme called the Manhattan Project where the atomic bomb was created. Nobody outside knew about it until after the

Hiroshima bombing. The town and some surrounding areas (code name 'Site Y') were created as a completely secret city, where scientists and engineers lived and worked at the Los Alamos National Laboratory. If you were even allowed to enter the gates, you had to avoid asking what townspeople did for a living. Thousands of people worked there, but no one outside could write to them without being subject to censorship. Los Alamos looks sinister from the surrounding hills – a sprawling network of hundreds of low-built, partly camouflaged buildings sitting on four rocky tables between the ridges of the Pajarito Plateau. I met a guy there named Jack who told me that he had moved there after headhunters had come looking for machinists at the Ford factory in Detroit where he worked. 'They said if I came to work there, I might help build something that could end the war,' he said. 'We didn't know for sure what that was.' But he found out on 16 July 1945, when he climbed a mountain to see the first atomic bomb test. He witnessed a mushroom cloud that was eight miles high. 'All of a sudden a flash went off,' he told me, 'and we saw the mushroom building up below us and then I heard the sound coming later.' Jack's friend Frank who also witnessed the test, took a photograph of the cloud; it's the only colour shot in existence. Not many people have watched an atomic bomb for pleasure, 'How did it feel – did it fill you with fear, or joy, or horror?' I asked. I was genuinely interested in their moral standpoint, witnessing such a horrifying creation. They explained that there was a degree of excitement from the staff, at the sheer power of the bomb. I asked: 'Was anybody troubled with guilt?' Frank told me that a large number of people were concerned, and that people in the town had circulated petitions – with hundreds of signatures – against ever using the bomb, but apparently they never reached President Roosevelt.

Despite the town's efforts, they had no control over their creation; which proves my point that the desire to become a politician should bar you for life from ever becoming one.

On a much smaller scale, I've enjoyed learning about some wee secret activities that take place in certain towns I've visited. For example, you wouldn't think gobbing would become a socially sanctioned activity in Edinburgh, Scotland, would you? But it did. Spitting on the Heart of Midlothian – a heart shape set into the cobblestones on the Royal Mile in Edinburgh – is a great old Scottish tradition. The Heart sits on the spot where people were hanged from the fifteenth-century Old Tolbooth. Originally, criminals would spit on the Heart to ward off fate, and even today you see Edinburghers spitting on it as they walk by. The Heart is not far from Parliament House, the gorgeous grey stone building with a columned facade that was once the site of the old Scottish parliament and now houses the Supreme Courts of Scotland. Next door is St Giles' Cathedral, the High Kirk of Scotland, and in front of it is a large bronze statue of a man wearing a lovely dress with angel wings. It's a monument to John Knox, who conducted the Reformation, the revolution that changed Scotland from a Catholic country to a Protestant one. That's when the dreary Protestant ethic of the Plymouth Brethren really got a foothold: 'Thou shalt not!' Were they kidding? This is a country where men wear skirts and no knickers. We fucking SHALL!

———

But once I get started on the age-old Catholic versus Protestant feud, with all the discrimination and prejudice that still abounds even today, you're not going to hear much else from me. Every Glaswegian schoolchild learned either, 'Catholics good, Protestants bad,' or the other way round. We were saturated with it growing up – including in the form of football 'team spirit', which was thinly disguised racial prejudice on both sides. It's been going on for centuries – and not just in Scotland.

Near Belfast, I visited a graveyard that was first allocated as a burial place by the government in 1869. There was a jostling for position. The local Catholic bishop insisted he got fifteen acres just for Catholics, and even demanded they build a wall to separate the dead Catholics from the dead Protestants. What did he think they were gonna get up to? Mixed marriages for the dead? 'Let's jump over the wall and shag a couple of Protestants, eh?' Fucking madness. Makes me so angry.

I'm fuelled by anger. And coffee. I used to drink coffee before going onstage and it helped to rev me up. But I didn't really need any help in that department. I get angry about anything that's remotely unfair. It's a great source of stage energy. All that anger comes raging up from somewhere inside and launches me into furious complaints and savage attacks. I can be absurdly angry – I've always been like that. It's been useful in my career, but it's never been fake or deliberate. I never had to summon it. It's just there, and it comes hurtling out with little provocation. Some people think I've got a computer tucked away somewhere with stories and rants on it. Really? I can barely work a computer.

During my *World Tour of Ireland* TV series I visited The Crown liquor saloon in Belfast. It's a brilliant traditional Victorian pub in Great Victoria Street, with the most ornate facade I've ever seen on a pub – Corinthian columns, and mosaic panels in butter-yellow, duck-egg blue and maroon. Inside it's full of wonderful touches – stained-glass windows, carved arches, low-walled wooden booths with leather seating, and an ornate burnished ceiling in gold and red. In the mid-nineteenth century, when it was called simply 'The Liquor Saloon', a couple bought it. He was a Catholic and she was Protestant. The woman insisted on renaming it 'The Crown' – against the guy's wishes. He went along with it, but he had his mates craft a picture of the British crown in mosaics and set it into the pavement in the doorway, so anybody coming in would walk all over it. That's Belfast political humour for you.

In Ireland and Northern Ireland, there are many instances of that kind of partisan shenanigans. During the great famine, the Church of Ireland would go round trying to get Catholics to convert. Their little trick was to give them food, which ensured that starving people would readily convert. But the converts knew what they were doing. They would feed up and get healthy – but once they got more flesh on their bones, they'd go back to Catholicism. In those days they were known as 'soupers and jumpers'. When I visited some starkly beautiful areas on the west coast of Ireland, Connemara gave me a special feeling of belonging because it's where my great-great-great-grandfather Charles Mills was born in 1796. He was a coastguard there, watching for pirates, smugglers, or people in trouble at sea. But he was also a Protestant! On the other hand, I also went to the Catholic graveyard where my great-great-grandfather Valentine Bartholomew was buried, with his wife Ann. My own wee genes must be raging war inside me.

———

I saw a different type of injustice, prejudice and persecution – at Pendle Hill in eastern Lancashire, England where, in the seventeenth century, communes of women with herbalist and traditional healing skills lived in forest regions. Two prominent 'witches' – Demdike and Chattox – and their communities fell afoul of prejudice at a terrible time of misunderstanding. Demdike's granddaughter Alizon Device was accused of killing a local pedlar by putting a curse on him, and the group was eventually accused of killing ten victims. Twelve people branded as witches – including two men – were marched forty-three miles to Lancaster Castle and all but one were hanged. I visited the grave of Alice Nutter and her family. She was hanged as a witch too, even though she was not part of their community, and there may have been some dirty tricks there. Alice was a wealthy woman, and it is now believed a magistrate who had been engaged in a fight with her about her land managed to lump her in with all the other 'witches' and stole her property. Bastard.

I also got angry when I visited the village of Tolpuddle in Dorset, which is considered the birthplace of trades unionism. It's where the famous Tolpuddle Martyrs used to meet under a tree on the lovely wide green. They were agricultural workers who formed a society with the intention of fighting against their landlords after their wages were cut. In 1834 George Loveless and five of his friends were sentenced to seven years in Australia, but they only served a year or so because there was an outcry of public support for their cause. They were pardoned and given farms in Essex, but local people treated them badly so most of them moved to Canada. I think they were heroes. Resentment based on political views seems to last for centuries.

Here's another place that inflamed my anger: the historic Kilmainham Gaol in Dublin, a forbidding-looking grey stone fortress with rounded corners and a vaulted, skylighted atrium that's now a museum. In 1916, Irish nationalist rebels who had besieged the General Post Office were brought there to be executed. The guards put white crosses over their hearts as targets for the firing squad and shot thirteen men. One of their leaders, James Connolly, was brought from hospital, sat in a chair and shot. I named my son James Connolly. The imposing building of the General Post Office was the headquarters for the Easter Rising against British rule in Ireland, which started there in 1916, and it's now one of the most famous buildings in Ireland. Patrick Pearse read out the Proclamation of the Irish Republic on the steps. The building was shelled from a gunboat in the river, and you can see the indentations in the original granite facade.

I learned about another terrible injustice that occurred in the seventeenth century: Oliver Plunkett, the Catholic Archbishop of Armagh, faced trumped-up charges of treason. It was obviously politically motivated, because even a Protestant jury found him not guilty. They sent him to England, where he was eventually found guilty and was hanged, drawn and quartered for 'promoting the Roman faith'. Some of his friends rescued his head and brought it first to Rome, and then to Armagh, and finally to Drogheda, where it now lies in St Peter's Cathedral. Plunkett was canonised in the seventies. When I was a wee boy, he was known as 'blessed' – the last of the Irish martyrs executed in England.

Once Australia became a British penal colony people were shipped off in their thousands – often for relatively minor crimes, like stealing a waistcoat. Some were even sent there for swearing. I'd be doing life. But there were far worse punishments in history than being sent to Australia. When Guy Fawkes was caught after trying to blow up parliament, they dragged him up the stairs of Big Ben in London and stuck his private parts between the hammer and the bell. It was only half past twelve, so he had to wait a whole half an hour.

———

What do poets do when they die? – Decompose. I once visited Dundee Loch – where the great William Topaz McGonagall sashayed around and composed some of his unique poems. He's one of my favourite people in Scottish history. Like all eccentrics, he was absolutely sincere in what he did. In his book *Poetic Gems* he writes, 'I don't like publicans. The first man to hit me with a plate of peas was a publican.' He was a wonderful nutter. He was an actor as well, although you wouldn't have wanted him in your play. He was once playing Macbeth and, because he thought the actor playing Macduff was jealous of him, he refused to die in the last act! A man who took defiance to the very limit! And who else on earth would have the balls to alter the Scottish Play?

Northumberland is a stunningly lush green rural county, and its large National Park has been designated as an Area of Outstanding Natural Beauty. I walked along a road built by the Romans, who were there for about five hundred years. They never really got on with the Scots, and they hated the northern weather. I mean, if you've come from Rome and you're wearing a wee white toga with short sleeves, the climate there isn't everything you would wish for. They eventually decided to give it a miss and built the impressive Hadrian's Wall – a 2,000-year-old wall with a straight edge! A lot of people think it was to keep out the Scots, but it wasn't really. It was a tollgate. People were allowed to come through to markets, with soldiers watching them. Two or three hundred years later the Romans fought the Picts – the real tough guys. All the tribes got together and kicked the Romans' toga-clad asses. At that point, the Romans must have thought: 'Who needs it?' And they came south. The Scots then continued to fight the English for a thousand years on and off . . . and it kind of continues. I don't know when it will end.

They seem to fight each other because they've always fought each other. It baffles me.

Some people believe Pontius Pilate was born in Perthshire to a Roman centurion who was stationed at Hadrian's Wall so, when I met David Bowie in Sydney where he was playing Pontius Pilate in the Scorsese film, I tried to talk him into playing him with a Scottish accent. To no avail. I think it would have been brilliant, though. He missed the chance of a lifetime there. The emperor Hadrian was a far more interesting man. He had a husband and a wife . . . and he built that wall. A bisexual brick-layer. Maybe he built it partly because he was a bit scared of the Scots. I don't blame him. I mean those Romans were wandering around in their wee togas and being set upon by blue naked people. They must have thought they were on the moon. They must have thought, 'Oh, okay! That'll do. We'll stop the Empire here. We've got to the edge of civilisation. It's the dark side of the moon from here on in.'

———

While some people give up, others keep the faith. In the church-yard of St Peter's Church, Bournemouth, Dorset, lies the heart of Percy Bysshe Shelley, although the rest of his body was buried in the Protestant Cemetery in Rome. The poet drowned in Italy at the age of thirty. His wife Mary Shelley, who wrote *Frankenstein*, was in England at the time. His body was retrieved from the beach by Lord Byron, Edward Trelawny and Leigh Hunt. Shelley was burned in a funeral pyre. Trelawny reached into it and grabbed his heart and sent it home to Mary, who kept it in a velvet bag round her neck for thirty years until she died.

Everybody has heard of Blarney Castle and the 'end of the

Is that a police car following me? Filming in the wilds of Arctic Canada.

Well, it's one for the money
Two for the snow
Three to get ready
Now go, babe, go
But don't you
Step on my sealskin shoes . . .

Look, Ma – no hands!
Yukon, Canada.

The Rambling Man's motoring fantasy.

Last one over the ridge is a Big Jessie.
Riding with cowboys in British Columbia.

Weird dancing pervert
spotted in the Pinnacles in
South Australia.

Australian artist Brett Whiteley's wonderful sculpture: the big matchsticks.

My meeting with the Walmatjarri Aboriginal artist, Jimmy Pike.

Not a whistling Dixie: Johannes O'rinda makes a living whistling classical music.

DIY brilliance: India Bhauti is a kinetic- and solar-powered busker in Australia.

Ramblin' Ralph McTell.

Ceilidh at home in the Highlands, dancing with Pamela.

The Highland gathering (Lonach march) with Sir Hamish Forbes.

I never knew the 'Devil's Rope' came in so many shapes and sizes. Visiting the barbed wire museum in McLean, Virginia.

Cadillac Ranch – inspirational 'Caddy' art in the
Texas panhandle. Graffiti artists welcome.

And my family
thinks my snow
globe collection is
over the top . . .!
Rob Lurvey and his
musical instrument
collection near
Springfield, Illinois.

Barney Smith's
Toilet Seat Museum,
Texas. My favourite
in this collection
was the *Jaws* toilet
seat . . . except you'd
have to put your
bum in its mouth . . .

A wee tune by Lake Pontchartrain in Louisiana, USA.

I've always liked graveyards.

Hank Williams' grave, complete with bowling green.

pier' idea that you're supposed to go there and kiss the stone. It has a distinct tang of bullshit about it, thinks not you? The real history is that in 1314 Robert the Bruce gave a wee gift to the man who lived there to thank him for his help in the Battle of Bannockburn when the Scots beat the English. The gift, which was just a piece of stone – a window ledge really – was sunk into a tower of the castle in 1446. Somehow, over centuries, it gained a legendary status. People started to believe that if you kissed this stone, you would immediately be endowed with 'the gift of the gab' – an ability to cleverly coax or flatter others. Apparently, this superpower was considered worth risking life and limb for, and to that end people would lower themselves in a dangerous fashion to kiss the stone. But after someone plummeted to his death while going upside down, they changed the rules. So now you have to lie down and bend backwards, while holding on. But that's not for me. It's not that I'm scared – I told you, I rather like heights – but I don't want to kiss a stone that's been kissed for 600 years by people with herpes and pyorrhea and all sorts of raging diseases. Yep. I avoided it like the plague.

11

THE SEXIEST LEGS IN LAUNCESTON

—

—

THERE WAS A Viking who used to own Orkney and his name was Magnus Barelegs. Oh YES!!! How come they didn't tell us about him in school? We were getting all that shite about Ethelred the Unready, and King Canute, and people in caves with fucking spiders – what about Magnus Barelegs – yippee!! I want to know more about HIM!!! There's a Glasgow line: 'I don't care if you're a Viking – get that fucking axe out of the dartboard!' Have you ever played with really good darts players? My father worked in a heavy darts pub in Glasgow. I said: 'You better watch it with those fucking animals. They might set about you some night . . . throw darts at you . . .' He replied: 'Just shows how much you know, Mr Smartarse. Darts players never fight. But *domino* players . . .!'

When I was in Little Havana in Miami, making *Billy Connolly's Ultimate World Tour*, I enjoyed playing dominos with the Cuban guys. They were lovely, sitting there in a little gated corner, wearing their panama hats. They played the game in that country way where you don't place the domino carefully, you slam it down. They go at a hell of a speed. Far too fast for me. They recognise the domino from a mile away and plonk the next one on the table while I'm trying to do the maths in my head:

'Hmmm . . . That's a four, but I need a three . . .' – he's got seven down by that time. It's great fun. They seem like very happy people, the Cubans who play it, a long way from the dangerous, domino-wielding hooligans my father warned me about.

I do have a history of playing dangerous games, though. When I was a boy, we had a daredevil game where I and other boys in my neighbourhood used to challenge each other to jump off the high air raid shelters in our street. We created obstacle courses called the Big Sui (short for suicide) and the Wee Sui, and there were plenty of accidents. A few years later, as an apprentice welder in the shipyards, I used to play a darkly dangerous prank with my pal Alex, electrocuting people when they stepped into puddles. One of us would stand near the puddle to look out for new victims, while the other would lurk below the deck with a welding rod, ready to zap. Even as a grown man I couldn't resist playing a very dangerous game of chicken with a bus driver in St Lucia.

But isn't a soupçon of danger what makes a game or sport interesting? While I was jumping to my possible demise from the Big Sui, other boys were playing marbles, which I found very boring. There just wasn't the excitement of imminent physical danger. But occasionally I joined in the marbles tournaments. We played on the surface of drains, which had metal coverings with little holes. You could flick your marble onto the surface until it settled, but if you knocked your friend's marble out of there you got to keep it. We swapped the marbles sometimes, but it was better to win them. Ronnie Meikle lived upstairs from us, and he had a beautiful marble that everybody admired. It was like a fish's eye. I won it from him, and he never forgave me. I bet if we met him today, he would mention it. Ya bastard, Connolly!

———

Rambling Men love playing tricks on people for a bit of fun. It's a way of communicating with all the new people they meet on their travels. Traditionally a Rambling Man carried a pack of cards so he could play a game with those he met on trains or in bars or pubs. However, if you play a Rambling Man and there's money on the table it's bad etiquette to win too much because, if you do, he'll be skint. You'd have to give him back the money just so he could exist. I've never been very interested in card games, and a poker face is well beyond me. But in my early hitchhiking days – and even later, when I was touring with Gerry Rafferty – I liked playing elaborate pranks on people.

Once, Gerry and I were playing Jimmy Logan's Metropole Theatre, and there were a lot of folk acts. We had to wait around for a long time before we went on, and, you know yourself, the devil makes work for idle hands. We were looking round backstage for things to do, and we found a fairly lifelike rubber snake. We tied this snake to a riser microphone that was sitting below the stage on a small mechanical platform, ready to rise vertically onto the stage. As the next singer – a female folkie – walked downstage towards the audience, the mic rose up for her . . . and just as she reached it, she saw the snake and screamed.

Then he'd sing an Irish song Bing Crosby recorded called 'How Are Things in Glocca Morra?' to nobody in the balcony. Well, one night Gerry and I crept upstairs and hid under the seats. At the end of his song, we suddenly appeared and shouted: 'Thanks very much!!' He jumped out of his skin. And there was another woman – Paddy Bell, who became a great friend of mine – who had to put up with our antics. When she was halfway through singing 'My Lady's a Wild Flying Dove' we threw a life-sized, fully feathered, rubber turkey onto the stage, which made her explode with laughter. Gerry and I eventually got fired for misbehaving – but we were reinstated a few weeks later, because the audience numbers had plummeted.

My cousin John was a legendary prankster. He was two or three years younger than me, but we went to the same school. He was a very intelligent guy, but he never liked rules. He hated the stupidity of petty rules. He always fought against the system and, like me, he was always in trouble. We instantly got on well. The first time I saw him at school, I was in for a pee, and I heard, 'Hello'. I looked up. There was a wall with a window in it and he was sitting on the window ledge. I said, 'Hello, John! How you doing?' 'Great.' At that time, the other guys called him 'Skin' because he was so thin. Great nickname. He came down from the window ledge and we talked and laughed. We made wee catapults from elastic bands, and we stood where boys were peeing in a line, and we fired orange peel at their willies. It was great fun.

When John and I were working in France together, he had a British Army wartime French phrasebook his father had given him. When we met women, we would consult the book and say things like: 'Is there any fluctuation in the price of eggs?' 'We have reason to believe there are Germans in your cellar . . .' The women would laugh and shake their heads, and that would break the ice. It was lovely. Those were really fun days, being silly in a goonish way.

When I was in my twenties, I visited John in England. He had a job in Bath, and he had a company van. That's where I learned he'd taken his pranks to a whole new level. He had tied a boxing glove to the end of a long stick, and he had written '*BIFF*' on the glove in white paint. We would be travelling through this park in the van and there would be lots of people riding bicycles. When one was approaching, John would say, 'Open your door!' He'd slide his own door open too, then he'd grab the stick and go up beside the cyclist. He'd say, 'Hey!' and the cyclist would go, 'Oh, hello' and he'd say: 'What's the time?' The cyclist would look bewildered: 'I dunno . . .' John would say, 'Oh fuck you then!' and shove the glove towards him. Bonk! He'd be flung into the bushes, and we'd be laughing hysterically. John and I remained pals until he died. We were loonies together. He liked to put people on edge. Did mad things like sticking plastic bullet holes on his glasses. He was incredibly funny in a bizarre way. People would think we were off our heads, but we were having a great time.

Some Rambling Men love tricks because they're entertainers at heart. They breeze through life and prefer not to take things too seriously. They join in with the game and aren't bothered if they lose. A true Rambling Man is also a thrill-seeker. He loves to throw himself into new experiences, especially if there's an element of danger to it. In fact, when it comes to following sport or games or dance, the element of danger is always what draws me to it. I love watching the ice dancing and figure skating on TV because it puts me on the edge of my seat. I can never decide if it's dance or sport, but it is both beautiful and brutal at the same time. I watch the skiing on TV sometimes. Boring as fuck. People in those awful suits skiing downhill all day. And then some big presenter in a blazer makes an important announce-ment: 'And now, the highlight of the day – the downhill race!' That's mystifying to me. Isn't that what they've been doing all day since dawn? I would have thought the highlight would be the *uphill* race. Then you'd see the guys with the *really* big thighs. Frog people.

Pamela used to take the kids skiing – not my cup of tea. Sliding on skis looks like great fun if you're good at it, but you go through years of humiliation to get there, and I've never been prepared to do that. Not this soldier. So, while she took our girls up the mountain to wheek down it, I'd just go along to watch people skating in the outdoor skating rink. I couldn't get enough of that. People crashing and falling on their arses every second. I laughed so much it hurt. I was intrigued by a thing Pamela said – that when she was learning to skate as a child she fell and cut her chin with the blade of her skates. So, God knows exactly how she had fallen.

One of the most barbarous and dangerous games I ever saw was in Igloolik. The Inuit teenagers play a 'Mouth Pulling Game' where they stand side by side, each with one hand behind their back, reaching for the inside of their partner's mouth, and try to pull their heads round behind them. People have been known to have their lips ripped off. And they play an extraordinary 'Kicking Game', where they jump and kick cans that are strung ten feet up. It's heart-stopping.

You wouldn't catch me doing any of that. I've not been much of an athletic type of guy – and I don't feel in any way ashamed of it. But I have had fun with a lot of the weird aspects of various sports: 'The Lord Provost said we're not having a marathon this year. It's going to be a half marathon. Why don't you have half a high jump while you're at it?' And don't get me started on my favourite sport at the Winter Olympics: the luge! That's the sport where you just lie down. Is there a sport where you do less? Just lying down, feet first, in the full-body condom? That's the sport for me. You train for it by playing dead. A luge champion's lying on his couch watching TV. 'Be quiet! Your father's training!' Cup of tea and a wee digestive.

I do like more gentle games as well, like flying a kite. It's one of the most sensual things you can do – fly a kite and sense the natural world on your wrist. Feel the energy of the wind coming up your arm. It's a lovely feeling. I taught my son Jamie to do it and he was amazed. I first did it with my sister Florence when I was a kid, in Kelvingrove Park in Glasgow. I made the kite with bits of wood and paper. In New York, next to The Coffee Shop in Union Square, there was a lovely man who sold wee kites. They were about the size of a panama hat, so you could fly them easily in the street about three feet from you. It was magical.

———

A great Rambling Man game is curling – at least it was the way I first saw it, on a frozen Scottish loch near Drymen, where I once lived. A local man took me, and I was amazed at how crazy it was. They were all drinking – there was whisky hidden in the leg of the brushes they used on the ice. Nowadays people involved in the sport wear spandex clothing but then everyone just dressed in warm sweaters and jackets that flapped around. The ice was cracking – *cccchhhccckkkkk* – but they couldn't have cared less. They were curling very aggressively and shouting and getting legless . . . it was brilliant. It was great to watch the raw game. I met a girl in Ayrshire who remains in my mind. I just started speaking to her on a train. She was really nice, and she was a curler; in fact she was in the Scottish curling team. I always remember her because for her day job she was a carpet-tufter. I thought that was a great job to have. When the Winter Olympics come round, I think about her. I wonder how my carpet-tufter is doing. I wonder, is there anything in carpet-tufting that makes you a good curler? Or in curling that makes you a good carpet-tufter?

I was never good at football or any of the things my father

would have admired. I love watching my football team Celtic play, though. Everyone around me knows I can't be interrupted on 'Soccer Saturday'. That's the day I refuse to attend to even minimal requirements of family life because I'm totally engrossed in the progress of my team. When they play their biggest rivals – Rangers – I have to watch the game lying down to avoid sliding to the floor with the emotion of it all. Yeah, Celtic are a great team. I wish I could say the same for some other teams I've seen. Eleven dickheads running out, half fucking drunk, all peely-wally, little wee legs. I have a tendency to shout at the TV. 'Jimmy! Pass it to me!' I'm talking to them like they're in the room with me: 'When you were in the hotel didn't you talk about the game? Didn't you have a wee plan?' We're the only team in the world that does a lap of disgrace. They get fucked and then they run round with their arms high, grinning. 'Yeahhh! Easy! Easy!'

I feel lucky I can get football on my TV – there was a time when Americans only broadcast American football, which bores me rigid. But there's nothing like going to a live game. I went to watch a Celtic vs Aberdeen game once with my pal Harry Frazer. He's a great Aberdeen supporter and we sat in his seats surrounded by guys in red and white striped everything. Naturally, they gave me a bad time, and it was great fun:

'Who's that prick in the green jersey?'

'Ah, it's the Big Yin.'

'What are you doing here? Don't want the likes of you around . . . you're gonnae jinx us, you are . . .'

I had a lovely time. People like to take the piss when they see me at a game. I was at a match one Saturday with my pal Huey Jordan and I was spotted very quickly by the crowd. One guy was showing off – he took out a ten-pound note very publicly and brought it over for me to sign. After I wrote my name on it, another wee guy came past, also making a meal of his request.

'Eh, Big Yin?'

'Yeah?'

'Will you sign my 10p?'

———

You know the ferocious dance you see at All Blacks rugby matches, the haka? Just outside Hamilton in New Zealand I went to a traditional Māori Pa (meeting house compound) and the tribespeople performed it – it is really an important ritualistic way of meeting visitors and deciding if they are friend or foe. After they perform the haka, you're supposed to respond in kind, so I sang 'I Belong to Glasgow':

> *I belong to Glasgow*
> *Dear old Glasgow town,*
> *Well what's the matter with Glasgow*
> *For it's goin' 'roon and 'roon*
> *I'm only a common old working chap*
> *As anyone here can see*
> *But when I get a couple of drinks on a Saturday*
> *Glasgow belongs to me.*

———

Despite the fact that it's clearly a working-class drinking song, the members of the tribe were under the mistaken belief that I was the king of Scotland. I told them I had a running battle with Sean Connery over that.

Back in Auckland a guy gave me a temporary *moko* – which

is a traditional Māori tattoo – on half my face. I loved the way it looked. But I was in the elevator in my hotel on the way back, and an American woman was standing next to me. She didn't seem to notice my *moko*, but halfway down she said: 'Bad hair day.' My life is one long bad hair day. People seem to think they can say whatever they like to me. I was in an Italian restaurant in New York minding my own business, and a woman just slid into my booth beside me and asked, 'What are you? Some kinda gypsy wizard?'

———

I had a lovely trip around the areas of the United States where zydeco music still exists. It developed from the French Creole or Cajun culture. The music combines Arcadian-French tunes with influences from blues and the Caribbean, played with guitar, washboard and accordions. When they sing it's a strange cross between French and American – and a little bit hard to understand. It's really a whole way of life – playing music, catching alligators, shooting 'critters' and clogging. But, that way of life is dying out. Employment is dwindling and I guess the alligators are winning.

———

I'm the type of Rambling Man who loves a good dance. In a nice little town in South Carolina is a shop that sells retro pots, pans and clothes during the day and then at 6 p.m. it stops selling the daytime trinkets and becomes a dance hall with old-time musicians and scores of people clog dancing. It's great. Exciting music that makes you want to join in. *Diggy do diggy do diggy do.* Feeling like there was no time like the present, I jumped onto the dance floor with everybody else and started clogging away. The whole thing reminded me of Scottish country ceilidhs, where people dance Scottish country dances and some-times take it in turns to recite poetry or sing. I especially love the dance 'Strip the Willow' when there are people of all ages in kilts just whirling around half-blootered, having a laugh. There's always a couple of people looking bewildered who have no idea what they're doing, with everybody else trying to prod them in the right direction.

Magnus Barelegs apparently wore a Scottish kilt – i.e. a short Gaelic tunic. He became the king of Norway in 1093 and was an illustrious Viking warrior who carried out many successful invasions of places like Orkney, the Hebrides and the Isle of Man. I can imagine he was dead proud of his legs. In that respect, I feel Magnus and I have something in common. In Tasmania, I won the 'Sexiest Legs in Launceston' competition. It was organ-ised by a crowd of nutters. Launceston had just won the national basketball championships and they were doing all kinds of crazy things, including that 'sexiest legs' contest, and, in a rash moment of bravado, I had signed up. I wore pink socks with pictures of Elvis on them – I think that's what swung it for me. I wore them without shoes, and just wore my underpants.

I not only have the sexiest legs, but my backside got voted 'Best Bum' by some Danish women in a pub in Copenhagen. I played folk music there for a few weeks solo after Gerry and I split up, and it was one of the sexiest places I ever played – a great city. The Danish women were all smoking yellow and blue pipes and they were very assertive. Afterwards, I vaguely remember going back with a woman to a kind of hippy place that had been an army barracks. The government had given up on it and the hippies had moved in. She offered me some kind of fruit that was hallucinogenic so, not surprisingly, I only have a hazy memory of the whole thing. But I think it was rather nice – the kind of thing Leonard Cohen wrote songs about. Anyway, you know what they say about the sixties – if you remember it, you weren't there.

———

There are loads of videos from TV shows of me running about in various exotic locations completely naked . . . except for my boots. I've danced bare-arsed in many parts of the world, but my maiden performance occurred in Cyprus when I was in the Parachute Regiment of the Territorial Army. It was my birthday, and tradition demanded that, after a few jars of local liquor to steel the nerves, the birthday boy had to strip naked, climb on a table, shove a lighted, rolled-up newspaper up his arse and dance until the flames had singed every hair on his bum then petered out due to lack of oxygen. Thus the 'Dance of the Seven Army Blankets' was born, and, I must say, it's a shame they don't do it on *Strictly*.

For reasons best known to myself I danced naked at the Ring of Brodgar on Orkney in Scotland. Why? I just felt like doing that – and the director of the BBC show wasn't going to stand in my way. I started off with a piece to camera, saying: 'These stones were placed here in 2500 BC by people we know nothing about, and for reasons we know nothing about. Was it a giants' board game? Were they symbols used in ancient numerology? Were they of religious significance? Or maybe . . . a landing strip for some intergalactic travellers? I personally think they're the goalposts for some ancient game, the rules of which are lost. It's just a very impressive place. And odd exotic hippy types from Glasgow have been known to come here and perform strange rites. Who knows why? We can but soak up the atmosphere . . .' Then I emerged naked and danced around the stones, singing, 'On Christmas Day in the Morning' to myself so I'd keep rhythm and help the editors.

Spurred on by the outpourings of appreciation I received after that, I dropped my trousers to dance round the Pinnacles in Nambung National Park in Western Australia. The Pinnacles are also mysterious structures. They are made of limestone and the same stuff as seashells and have been shaped by wind and sea over many centuries. In recent years, New Age people have begun to gravitate there to find a bit of peace. Again, I faced the camera with what seemed at first to be a serious explanation: 'Wherever you get phenomena like this in the world – like the Ring of Brodgar in Scotland or Stonehenge in England – there's usually some strange character who is drawn to such things . . . a curious person lurking about who is usually a bit of a freak . . .' Then I emerged and did one of my naked dances all around the Pinnacles.

My most uncomfortable naked dance was when I danced in the snow in the Arctic Circle when I was filming *A Scot in the Arctic*. It was terrifyingly cold. I kept my sealskin boots on, but I hoped to have the future use of my limbs and willy, so I kept it very brief. When the show was broadcast, the editors put an animated snowflake on my willy to protect the morals of sensitive viewers. Apparently, my bum can be on display no problem, though. Unlikely to offend or overexcite viewers. I beg to differ. Not many people know I'm Tom Cruise's bum double.

———

Even though my bum's been seen by millions of people all over the world, I haven't seen it that many times myself for obvious reasons. But I was once shown into a hotel room and was both surprised and horrified to see there was a mirror above my bed. I thought, 'Oh no. Oh God. What's expected of me now?' I tried out a few positions – me on top, her on top – trying to figure out the angles. I was on my own at the time. Pamela was in the bathroom getting ready. I was practising so I wouldn't disgrace myself. You so rarely see yourself lying down. I was all spread out. And I thought, 'Oh fuck! I've become my mother!' It was the weirdest thing. I've never understood why someone would be wanting to be shagging away and look up and see a big white spotty arse. I thought maybe I should write jokes on it or something.

The sex life of a Rambling Man is frankly no one else's business. But I'll reveal to you, from my personal experience, that I've been lucky enough in my short but colourful life so far to have been the recipient of a blowjob or two. And to the best of my memory there wasn't a lot of blowing going on. It just seems, if it comes to the Trade Descriptions Act, that it's wrongly named. People could get hurt you know. They could get wounded. I mean if someone is going to give you a blowjob, but they've never done it, they've just heard the name . . .

'Any danger of a wee blowjob?'

'Oh certainly . . .'

Whoooooo . . .

'What do we do now? Sing "Happy Birthday" and make a wish?'

Or even worse, they might blow up my willy like a balloon until it bursts. *BANG!* 'Oh Jesus. Stay where you are. I'm gonna put the light on – my bollocks are on the carpet here.' Scrotum hanging like a wet chamois.

The scrotum is such an awful-looking thing, don't you think? It's an ugly piece of stuff. Like a hairy brain. It's a design fault, isn't it? I have never felt another man's scrotum, but I imagine your first touch of a scrotum is the most unpleasant thing on Earth. 'What the fuck is this? Something very unpleasant has just crawled out of your arse!' No wonder they keep it tucked away in a corner. When bodies were being designed . . .

'Er, God?'

'What is it?'

'We've got a lot of elbow skin left . . . what are we gonna do with it?'

'Make wee bags. We'll put their balls in them. Save them carrying them around in their hands.'

But it didn't work, did it? Men just love touching their balls, as if they are in constant need of adjustment. For some guys, you barely notice the action. They can be discreet – touch them through their pockets. But others just openly hang on to them. They don't care. They even do it while they're staring at you. 'How you doing? Yeah, I'm just holding on to my balls. What's happening?'

'Scrotum' is a terrible word too, don't you think? One of the worst words you'll ever hear. 'Penis' isn't much better. Now, 'vagina' is a lovely word. Sounds like a nice place . . . and it bloody is too. If you saw that on the back of your cereal packet: *'Two weeks in Vagina could be yours! Enter this simple contest,'* you would say, 'I think we'll go for that!' Right? Although, I was passing a school in Sydney once and some kids ran out to have their T-shirts signed. One boy said, 'Mr Connolly, I think you're a great comedian. I love that bit where you say, "Vagina's a lovely word."'

What have I done to Australia? He was about twelve.

Maybe I'm too friendly to strangers. I'm an average happy

type of guy. I always say 'hello' to people. For a second, they think they know you but they've forgotten who you are.

'Hello! How've you been?'

'I've been fine.' You can see them trying to work out who you are.

'Haven't seen you for a while . . .'

'Och no – I'm here on tour again. I'm just here playing the Opera House . . .'

'Aghhhh . . .!!!'

———

Sex can be a very frightening thing. I remember my first sexual experience. I was very frightened because it was dark and I was all alone. But masturbation is a healthy thing. Don't knock it. I fainted in church once on a Sunday from having an erection for nine days. See – abstention can be very bad for you. Masturbating is the only exercise some people get. You don't have to feel guilty. No one's monitoring – I've met religious types who tell you there's people flying in the sky to tell God you're wanking.

'Oh look! There's someone doing bad things! We'll tell God about him.'

'God?'

'What is it?'

'Someone's wanking.'

'Who's wanking?'

'Billy.'

'Oh, he's always wanking.'

'God – he's wanking again!'

'I KNOW!!'

You think God pays attention to who's wanking? I think he's pretty busy in Ukraine. Personally, I do it in the morning to get my heart started, for medical reasons. Plus, it has other advantages: you don't have to look your best. You don't even have to shave or wash. It's a wee present from God for the lonely. I'll never forget my most extraordinary masturbation. I was pounding away there. I had the magazine open in front of me. Bashing away. And the man in the shop said, 'Are you gonna fucking buy that?'

I was in a museum of torture in Melbourne and, of all the nightmarishly horrendous punishments, I witnessed devilishly cunning big circular leather things. They were designed to stop self-abuse among the prisoners. In other words, they were anti-masturbation gloves. Mean-spirited bastards. The least they can do is let you have a wank in jail.

There was a famous highway robber called Ned Kelly who was jailed – Australia's Robin Hood. They have his suit of armour in the museum, and a wonderful letter he wrote before his execution:

I have been wronged . . . no alternative . . . but to put up with the brutal and cowardly conduct of a parcel of big, ugly, fat-necked, wombat-headed, big-bellied, magpie-legged, narrow-hipped, splay-footed sons of Irish bailiffs of English landlords.

———

Go for it, Ned. I love it. His last words were 'Such is life'. I love it when men write or speak powerful words that last through the ages. Robert the Bruce – a man I admire not only for leading Scotland during the First Scottish War of Independence, but also for not fearing spiders – signed the Declaration of Arbroath, in which he and others declared that, as long as 100 of them remained alive, they would never bow beneath the yoke of English domination. Stirring stuff.

I lost my virginity in Arbroath, Scotland during a camping trip when I was a teenager. To my eternal discredit, I didn't know her name then and I don't know it now. She was very nice to me, though. It was in Wardykes Campsite, next to the graveyard. I looked for it when I was touring there – it was like looking for the source of the Nile. Somebody lives over the spot now, and they'll no need central heating there let me tell you. The hit at the time was Connie Francis's 'Lipstick on Your Collar'. The blood was draining into my willy. Shame there are houses on the field where we camped now. They should put up a statue. Some kind of erection. It was such a momentous part of my youth. I can see my old motorbike standing there – a BSA C12 overhead valve, 250cc. Brilliant.

That reminded me of a girl I used to know in Glasgow. I was about nineteen. She was a jolly person . . . and she had one of those flats in Pollokshields in a red sandstone building with a bay window. The bed was in a recess alcove and really high up, with a spiral staircase leading to it. It was at a period in my life when my pals were lying very heavily to me. Male teenagers can be so stupid. They'll say, 'Oh have you tried it with your leg over your shoulder and you put your hand here . . . she slides under, she bites your knee, and you lift . . .?' Nonsense. But you believe it all. 'Fuck. I've never done that. Must be something wrong with me.' So, one night I was in bed with her late at night and we were wrestling about doing normal stuff – *kissy kissy heavy sweating* and the lamplight was shining in the window *heavy breathing sweaty sweaty* and it was nice. I said, 'Listen, I've got this idea . . .' I actually had no fucking idea. I was trying to remember what the guys had said to me. I knelt up. I said, 'Give us your leg . . . wait a minute, I think it's . . . no, it's this . . . no hang on – come a wee bit nearer . . . oh fuck, I've got cramp – move! Move! . . . Oh fuck, my back!'

I'm doing all this trying to get near enough to do something . . . 'Just a sec.' And I knelt on what I thought was the edge of the bed – which proved to be a shadow. My knee shot in the direction of hell, and my balls hit the edge of the bed – the metal bit. *Bong!* FUCK! My pupils crossed each other *WHACK!* My left leg came right over my head. I came out of the bed like a helicopter in trouble *AAAGGHH.* I plummeted to the floor. *CRASH.* I landed in a bundle. A bookcase fell on top of me. And she leaned over and said: 'Brilliant. What do I do?'

———

When I was a younger Rambling Man, on the road in my folk-singer days, I would often sleep on someone's floor. Sometimes they'd give me a bed, but it would have cheap nylon sheets on it. You *cannot* have sex on nylon sheets. You can't get any purchase. You slip all over the place. 'Get off of me!!' Your pubic hair stands up and sparks come off it. If you've got a wee jaggedy edge on your toe, you catch it and pull the whole sheet off.

Don't you find sex awful complicated? It used to be so easy – bash, bash, then away for a kip. Then things started to become complicated. Women started making demands. Too much pressure leads to: 'Sorry, it's never happened before!' followed by: 'Billy . . . I'm married to you . . .'

Now, you're probably thinking that windswept and interesting showbusiness Rambling Men like myself must have shagged a different groupie every night back in the day, but it wasn't like that, honest. Sex on the road is nothing. It's like doughnuts. Not a great thing. Always ends up being kind of nasty. Grubby. Some guys are kind of addicted to it. You can feed any habit on the road. I met some of those girls who are part of a group known as the Chicago Plaster Casters. Everyone in rock 'n' roll knows who they are, and they were featured in that movie I did with Bill Nighy – *Still Crazy*. Cynthia Albritton started it. She was a visual artist who came up with the idea and developed the technique of making plaster casts of erect penises. Jimi Hendrix was the first well-known person who agreed to take part. Her assistants in Los Angeles – and similar groups in other cities – became famous. They were sexy women who knew everybody – all the head roadies, all the big money guys. So, they'd meet a guy who was part of a band – preferably the lead guitarist – and they'd make him an offer he couldn't refuse. Some guys would willingly go along with it to be immortalised in that way. And guys who'd had it done were well proud of it.

I had a friend called Booby Daniels who was the ultimate roadie . . . and the ultimate object of feminist disdain. He'd do anything to get laid. He'd be out all night working on it. His room was a disgrace. I called that look 'the exploding suitcase'. He was a wonderful guy. He was from Glasgow and his mother had died tragically after falling out of a tenement building when she was washing the windows. I met him when he was working with Elton John and Ray Cooper and he took a shine to me. I used to make him laugh. He was very funny and very generous. He was successful in his efforts to get laid, even though he wasn't attractive in the conventional way. In fact, he looked like his face had been squashed. He addressed every woman he met by one name. It may have been 'Lisa' – I can't remember. Sometimes it was a successful ploy. I remember it working on a waitress in the southern states. He said, 'Hello, Lisa!' and she said, 'Hello – how do you know me?' He said, 'I know all sorts of things about you.' He started talking to her. Next thing he was leaving with her. He never married, but about a year before he died he found out he had a daughter. She came to visit him. She loved him, and she was at his bedside when he died.

———

During the course of my film career, I've had to get into bed with beautiful, famous actresses. It's been hell. And before you accuse me of being a show-off, I need to explain that sex scenes for movies are ridiculous. You're doing all this sexy acting and there's forty people in the room, reading the paper, eating chips, looking out the window . . . and it's roasting. The director comes over . . . 'Could you be more vocal?'

So, I'm trying my best. 'Oughh . . .'

She's going 'Yes! Yes!'

I'm thinking, 'I don't remember asking you anything . . .'

See, in my real sex life, I've got fucking nothing to say. I'm just grateful. Anyway, I have hardly any voice left after the three hours of pleading. I did a movie with Sharon Stone, and we weren't getting along. During the bedroom scene the director came over to me.

'You okay, Billy? Thought it might be awkward being in bed with someone who's not talking to you.'

'You fucking kidding? I've been married for a long, long time.'

Sex should be fun. It's a hilarious thing in many ways. Especially when you're off your face and you watch pornography backwards. People going instead of coming. It's very weird. The willy becomes a sort of hoover *Thum thum thum thum.* People jump off each other and into their trousers. The whole thing ends with everybody fully dressed round a coffee table talking.

You should be brave and experimental when you have sex. I saw a birthday card in Los Angeles – it was a big he-man holding a birthday cake across his willy with the candles all lit. It said: '*So it's your birthday . . . Fuck it!*' I've never done that, but I have rubbed my willy on a fondant fancy. In a shop – it wasnae mine.

12

PLAYING BANJO WITH THE LESBIAN MERMAIDS

—

IN THE LATE sixties, it became fashionable to be a certain type of Rambling Man – a hippy version. When I was still a welder, I was watching a programme on the BBC called *Tonight* with Cliff Michelmore and they had a segment about a phenomenon happening in St Ives in Cornwall. All these hairy people had moved there living in caravans and cottages. They had formed a community and were playing music and having a blissful time. I was totally taken by it, especially by one of their leaders, Wizz Jones, a singer-songwriter who had a beard and hair halfway down his back and was playing the blues really well. When I met him years later and told him that he was such an inspiration for me he just laughed. He said, 'I got letters from everywhere, from people complaining that their sons were badly influenced by me.' So, I wasn't the only one saying to myself, 'How can I get to look like that and play like that?' He was an amazing guitar player. It was great to see people doing what I dreamed of doing, like Alex Campbell, Ralph McTell, and Clive Palmer from the Incredible String Band. They were busking, and sometimes doing the clubs – campaigning against steady jobs. They had firmly decided to follow a different kind of lifestyle to most other people, and I really admired that. I wanted it for myself.

Bert Jansch was another Rambling Man I admired at that time. He wrote great songs, and became internationally famous, with dedicated fans all over the world. He was a friend of Ralph McTell, and I met him in the sixties in a folk club. He was a loner, and not very comfortable in the company of others. Women loved him and I did too. He was *thrawn* – that's the Scottish word for 'gaunt'. His cheeks stood out. Thrawn's a better word. He once told me something about himself that just floored me. He said that he'd heard the great guitarist Davey Graham had been in North Africa – in Morocco – playing guitar and picking up on the local Arabic music. So, Bert decided he would like to go to Morocco too. He went, and he wore a duffle coat the whole time he was there. That was all that was in the story. I just fell off my chair. He was a Rambling Man folk singer in a duffle coat in Morocco. He was so naive. He said: 'I was roasting, but I didn't want to take off my duffle coat. It's a good coat.' I met Bert for the last time in Aberdeen. I saw him on the pavement. He was playing at a place called the Lemon Tree, and he said: 'I just got married.' I said, 'Are you kidding me?' He said, 'Yeah. This is my wife. We got married in the Isle of Arran.' If you attend a concert of any folk singer-songwriter, there will be at least one song about love on the road. 'I met her on the road.' Or 'I miss her . . . I was on the road.' It's a distinct lifestyle.

Lots of guys I've met who were in bands fell into the Rambling Man category. Being in a band is a Rambling Man subset – going on tour and doing the thing your parents knew nothing about, and didn't understand. The vast majority of folk singers were Rambling Men. They'll sing about the ability to get away from their troubles by finding freedom on the road; find other people who feel the same, and stop being miserable. Many of Bob Dylan's songs are about being on the road and meeting people on the road, being disappointed in love on the road . . .

> *When your rooster crows at the break of dawn*
> *Look out your window and I'll be gone*
> *You're the reason I'm a-traveling on*

———

It's the same with Leonard Cohen. His song 'Suzanne' is a classic Rambling Man song about a guy who found somewhere to stay the night – with a female Rambling Man.

> *Suzanne takes you down to her place near the*
>   *river*
> *You can hear the boats go by, you can spend the*
>   *night beside her*
> *And you know that she's half crazy but that's why*
>   *you want to be there*

———

People who are purely songwriters are not usually Rambling Men; generating a living writing songs is a whole different thing. Real Rambling Men may want to write songs, but we don't necessarily want to earn a living that way. Our songs are just precious expressions of our lives that are not primarily designed to be commercial.

Lots of people in rock 'n' roll are Rambling Men. Not so much the stars – they've got a job that takes them on tour and then they go home. But the crew often go from one long tour with one band to another with a different band. They've chosen this rambling way of life. Some don't have homes; they just stay with friends or family when they go back to their hometowns. There are some Rambling Men with big names though, like Davey Johnstone, who played guitar in Elton John's band. I met him when we were younger. We played in clubs with the Incredible String Band and slept on people's floors. It was lovely. In typical Rambling Man fashion, we didn't hang around one place for too long.

The minute I met Ralph McTell I knew he was a fellow Rambling Man. He played acoustic guitar and sang wonderful songs about searching, longing and discovery. I became his friend, which was a wonderful thing – being a friend of a guy who was the real deal. We understood each other. Ralph wrote and sang the marvellous song 'Streets of London', which is a kind of Rambling Man anthem.

The romantic nature of the Rambling Man became part of the zeitgeist in the sixties and seventies. There was a longing attached to it, and sometimes – as in Tom Paxton's lyrics for 'Ramblin' Boy' – a sense of staunch loyalty, camaraderie and loss:

> *He was a man and a friend always.*
> *He stuck with me in the bad old days,*
> *He never cared if I had no dough,*
> *We rambled around in the rain and snow.*

———

Tom Waits was the king of the showbiz Rambling Men, and Woody Guthrie was the emperor. They even made it a lucrative lifestyle, while managing to maintain an image of being just simple rambling folkies. We were all on the periphery of it – I was far from being a Woody Guthrie in my folk-singing days.

I had loved American country music since my childhood, and even though many of the really big country music stars like Glen Campbell, Johnny Cash, Dolly Parton and Bobby Vee hardly seemed like Rambling Men because what we knew of their lives was fed through well-oiled career operations and all about *People* magazine, mansions, red carpets and the Grand Ole Opry, but I and other Rambling Men enjoyed their music, because they sang about life and disappointment and joy – real things. They did sometimes take the tragic element a bit too far, which is why I've taken the piss out of country singers quite a bit. I've even created a recipe for writing a truly tragic country song. What you need is relatives – the closer the better, so your mother's great – get her in there. And God should put in an appearance, plus a disaster or two, as many as you like. You should have somebody in it who has something terribly wrong with them – preferably a terminal disease. I wrote a country and western song that was in very, very bad taste. As a matter of fact, it was sick. It was called 'My Mother Drowned in the Grotto at Lourdes because a Hunchback Pushed Her In'. I took it to my publisher, and he said, 'You'll never get away with that, Bill, it's a disgrace. You need treatment, Bill. It's offensive.' So I cleaned it up, and it's now called 'How Can I Tell You I Love You When You're Sitting on My Face?'

When Rambling singers sang about love it wasn't about lasting monogamous relationships. None of that *'Goin' to the chapel and we're gonna get married . . .'* It was never about marriage – that all ended. Rambling Man songs were still about love, but more often about temporary, brief experiences. They were about freedom, and longing, and meeting many different people along the way whom you could love for a while. But not for ever. The social landscape was changing in tandem with this movement. You could even wear each other's clothes. I once swapped trousers with a guy on a bridge in Amsterdam. Perfectly normal. See, early on, looking good was important. Wearing the right jeans was important – like Levi's 501s. I might have been a bit too fashion-forward for my own good; when I first wore flares, people called me Popeye. We rambling folk wore desert boots, cowboy boots, denim and cheesecloth shirts. Our outerwear was donkey jackets, motorcycle leathers, corduroy jackets, and we flung on scarves and beads. I used to dab sandalwood oil on myself, and I wore patchouli oil for a while until I got flung out of a pub in Quebec for wearing it. 'You smell like dogs' piss.'

In the sixties and seventies the guitar you played was also important. There were fashionable guitars and unfashionable ones. You'd hear someone saying: 'See, so-and-so's okay, but he plays a Welsh guitar.' Snobbery. The best guitar to have was a Martin. Gerry Rafferty lost my guitar on a train, and he has yet to apologise. It was a Harmony Sovereign – great guitar. At the time, Gerry didn't have a guitar . . . well, he had a guitar, but it had been painted by the artist John Byrne so Gerry didn't want to take it on the road. Out of extreme kindness, I lent him my guitar, and he left it on a train. Bastard.

Ukuleles were not considered fashionable. Those who had them were seen as just plinky-plonky players who couldn't express the same kinds of high emotions as guitarists – love, regret and longing. We considered them to be the same as people who played bongos – pains in the arse. Mind you, lots of people feel the same about banjo players. My pal Eric Idle once commented: 'Banjo – the musical choice of the antisocial.' He's right, and from the beginning of my fascination with the banjo I knew it was not always a popular instrument, but I didn't care. Since I was a boy, I've been in love with the noise it makes. I've played the guitar as well, I've dabbled in the ukulele, and I love my autoharp, which is an easily commanded instrument that sounds lovely with the acoustic voice. Nowadays it's hard to play my banjo because, due to my Parkinson's disease, my left hand won't do what my brain tells it to do, so I've started playing my harmonica more and more. I really enjoy it, especially for playing the blues. The harmonica is big in the Rambling Man world. It has a plaintive sound, and you can just stick it in your pocket.

———

Playing music is key for a lot of Rambling Men, because it's an artistic form of expression that you can take wherever you go. You have to choose your instrument wisely; there's a reason why there aren't any cello-playing Rambling Men. I've collected some nice instruments over the years, but my own collection pales into insignificance next to a musical instrument collection I saw near Springfield, Illinois. The owner, a man called Rob, had the most impressive, eclectic assortment of musical instruments I could even have imagined, including the largest collection of lap steels in the world. He had just under two thousand electric guitars, and even collected zithers and ukuleles. The man had all this stuff displayed on his walls in a massive, secret space. He was not a dealer or anything – it was just his private collection. He never sold anything. He told me he collects other random items, like tweed suitcases and medicine bottles. He calls himself an obsessive-compulsive, manic-depressive, eccentric collector. I love that he and his collection exist in the world.

My love of folky-type musical instruments goes back to my boyhood and my heritage. Part of me is Irish. One side of my family – the Connollys – comes from the west coast of Ireland and the other side – the MacLeans – comes from the west coast of Scotland. Whenever I go to these coastal parts of Ireland and Scotland, I get a wee tingling. I feel it's where I belong. Our culture is wedded to music. In Ireland there are lots of gatherings of people who just get together privately, or in pubs or halls, to sing and play music for its own sake. They are great at it, and it keeps traditional music alive – the stuff you don't hear on the radio. But my favourite place to play informal music, just sitting round in the pub jamming with friends, is on the Isle of Arran in Scotland. Over the years, I met lots of pals in Arran that I'd met in the folk scene and through camping there, and a whole group of them eventually moved there to settle down. So, whenever I visited Arran I played with the whole gang – Geordie who was a banjo player, Zoe who was a great singer, Johnny who played the tuba, concertina and pennywhistle . . . and several others. We played together in the local pub next to the village hall, and it has always been a terrific pleasure.

Those Arran friends fully embraced the Rambling Man style, but not all my old friends could see the point of it. I used to go and have a beer with the guys in Partick who were unemployed, or who worked in the shipyards. They would make jokes about me after I'd been away performing. They recognised that I was different . . . but I recognised that they were the same. I had simply made a choice to live in a different way and appear in a certain way. Sometimes that can be confusing for people. There was no rulebook. The difference between us was just our understanding of success. To those who didn't understand what we were doing, success meant getting a certain type of job or achieving a degree of financial security, but to Rambling Men your most coveted type of success is gained through the strength of your spirit and richness of experience in the world. It was about being what you wanted to be rather than what you thought you should be. Our heroes could range from people like the folk singer-songwriter Hamish Imlach – who lived in a council house with a family, but was one of us – to the famous guys, the big folkies: Wizz Jones . . . Clive Palmer. They topped the bill at a club then went to sleep on someone's floor.

My all-time country music hero Hank Williams, who sang 'Long Gone Lonesome Blues', is buried in Birmingham, Alabama. He was The Man. His music was extremely important to me when I was a boy. There was a border of stones round his grave. You're not supposed to take one, but I couldn't resist. Hank lived most of his life there in Birmingham, but he ended up in Nashville as a huge international star with eleven number-one hit songs. He died when he was only twenty-seven, from doing alcohol and drugs in the back of his car on the way to a gig. 'Let that be a lesson to you, m'boy.' That's what my father said about it.

———

I love being in places where I can hear extraordinary, unusual live music, like the band they had in Texas for a high school football game. There were 350 people in the band, and twelve xylophones! I'd never seen twelve xylophones together before. They sounded like a brass band and were extremely good. And there were 300 people in the choir – for a football match! It was very different from football matches back home: no pies, no

Bovril, no chanting of taunts regarding a personal sexual act aimed at the referee. A huge, highly animated crowd attended. A woman was trying to get me excited, but I didn't understand the game. Story of my life.

The best live music I ever heard while on the road was Bob Dylan at a concert hall in Canada. He didn't look at the audience. He played his piano facing the side of the stage and never turning round to face the crowd. You never know what you're going to get with him. He was brilliant. The ultimate Rambling Man. Almost all his songs are about being a Rambling Man, with true open-road wandering lines such as: *'Where I'm bound, I can't tell'* and *'I'm a-thinking and a-wonderin' walking down the road'.* Yes, he never wanted to stay put.

I earned money as a busker from time to time when I was hitchhiking around France, before I could get proper gigs as a folk singer, so I know it takes extraordinary courage to be a street performer. Your audience will always be terrifyingly unpredictable. Since then, I've always loved coming across great street musicians, especially if they are doing something I've never seen before. I found some excellent buskers in Australia, like India Bhauti, the one-man band I came across in Sydney. It was a joy to see him playing weird electronic drum music with a solar panel on his head that powered his handmade instruments. And Johannes O'rinda, who performs extraordinary feats of full-body coordination by whistling classical music while conducting with his arms at the same time. Then there was the sensational Kokatahi Band, playing 'bush music' with a line-up of accordion, fiddle, banjo, drums, spoons and more.

There was a lovely moment in an Inuit town when I drove up on my skidoo and witnessed something extraordinary. There was a girl singing in a very unusual way. She was facing another girl who had her mouth half-open and was making shapes with it, so the first girl could project her voice into the other girl's mouth and the echo created different noises. I learned this was throat singing, and it was the most unusual sound. I came up on my bike and I said, 'How are you doing that? It's amazing!' Then a head came round from the first girl's head and I realised she had a baby in the back of her anorak. I said, 'I know! *He's* doing it!' I was blown away by that style of music-making, and it's still a mystery to me. It was a rare and beautiful sound they made.

I want to tell you about the most incredible impromptu musical performance I ever witnessed. It was a private one – only I and one other person saw it – and it remains a highlight of my life. Remember I told you about the first time I was in Malta, after the plane got hit by lightning and we had to land there for repairs? Well, my second trip to Malta happened almost as suddenly. Totally unplanned. My manager at the time was going there for a couple of days to do some business. He had clients filming on location in Malta for the movie *Popeye* starring Robin Williams. 'You wanna come along for the ride?' he asked. 'We'll have a laugh.' 'Sure.' I'd met Robin Williams beforehand, in Canada on a TV show. He'd been funny about my green leopard-print suit that he called my 'Irish Tiger'. My manager and I went along to the set of *Popeye* and Robin came into the room and looked at me like his memory was whirring, and then he said, 'Canada!' I said, 'Yeah!' See when I first met him in Canada, he was very famous, but I didn't know who he was. 'What do you do?' I had asked. 'I'm an actor.' He was lovely. We got on very well right away.

That first night I was in Malta, we all met up – Me, Robin, Ray Cooper the percussionist, a genius banjo player called Doug Dillard, and Harry Nilsson, the brilliant composer and singer.

I'd loved his music for years. He was doing the music for the movie.

The plan for that night was for us all to get shit-faced. We started drinking and it was all very jolly, and then Harry said to me, 'Before you can become one of the gang – a Member of the Knights of the Maltese Cross – you have to write your name on that castle where everybody can see it!' He pointed to a huge, towering, limestone fort sticking out of the landscape up a steep hill. Everybody looked at me as though they were thinking, 'Surely he's not going to fall for that?' But, undaunted, I climbed the hill and scaled the tower and wrote *BILLY* on it in large white letters with chalky white stones I found lying around.

Primary mission accomplished (getting shit-faced), we ended up in a nightclub. That's where we got into a fight. The legendary roadie Booby Daniels had showed up, and – true to form – he was chatting a woman up next to me. On her other side there was a Maltese guy and, after a while, he mumbled something to Booby, and Booby mumbled something back, and then the Maltese guy hit him in the head with an ashtray. It was a real cowboy fight – people walking backwards, kicking. We all exited the club at high speed. Outside, after taking another one on the chin, Booby stumbled and knocked down the marquee. After that, everybody dispersed for the night. We dropped Booby off to attend to his injuries and likely concussion, and then it was only me and Harry left. Harry said: 'There's a guy here who plays great guitar. He works in a garage.' Well, we found this guitarist on the way back to our hotel, but he hadn't brought his guitar. There was a piano sitting there – a terrible mess of a piano. It was painted green. So, Harry sat on a bar stool above the keys, fiddled around with the instrument for a bit, and then he turned to me and said: 'What do you want to hear?' I said,

"'Remember Christmas'". I love that song of his. Harry played it, just for me and the guy in the garage, and it was one of the best moments of my life. Not many of them in a pound.

———

I like Malta. I like its unique sites, such as the prehistoric Ġgantija or Gargantuan Temples in Gozo, which pre-date the pyramids of Egypt – they're the oldest freestanding structure in the world. The people in those days were obviously giants. Big doors, huge steps – those people must have been enormous. They might even have been big aliens. It's a lovely site, and when you visit you can touch things. They don't care – they don't have notices everywhere telling you to keep your hands to yourself.

My favourite cave in the world is Ninu's Cave in Xaghra, Gozo. It's a wee cave that was discovered when a man called Joseph Rapa was digging a well under his house. He kept quiet about his find for a long time, because he was afraid the government might commandeer his home. Now it's a wee tourist attraction that people like because it is so uncommercial. You enter it through the front door of the house. You knock, and, when someone answers, you say: 'I'd like to see the cave please.' 'Certainly. This way.' You edge your way past family members who are just going about their business in the house – cooking dinner, watching football on the telly and taking choir practice for the local church – which is the best bit of the whole visit. Next to the bathroom is a door that leads to the wee cave. You climb down some stairs, and there in front of you are some limestone formations to which the family has given names like 'The Elephant' and 'The Organ'. You can spend a few moments trying to see the resemblance, but pretty soon a family member will let you know your time is up. You put a couple of euro in the box, and you're ushered out the door. No T-shirts, no gift shop, no hard sell of knick-knacks. Brilliant.

I also went into Calypso's Cave in Gozo . . . it's supposed to be where Ulysses showed up during his travels and fell in love with the nymph Calypso. I mean, Ulysses was the original rock star, wasn't he? Going off on tour and swishing round the world . . . with the long-suffering wife Penelope sitting at home weaving her tapestry, and Calypso being the beautiful groupie that he fell for and lived with for years, right? I'm all for it, but don't mention that to Pamela.

———

While I was touring Australia once, I became a proud member of a fake rock band invented by Davey Johnstone and me in Perth. It was a rock band called Chilli Baby Squid and the Lesbian Mermaids. Davey Johnstone and I were in a seafood restaurant and the menu was decorated with elaborate drawings, with each special dish described in a whole paragraph, separated by a drawing of a mermaid. I said, 'I think I'll have the chilli baby squid.' Davey said, 'I know him.' I said, 'I know him as well. He lives in LA. He has a band called Chilli Baby Squid and the Lesbian Mermaids.' We invented this whole history for them, and even had T-shirts made with CREW on the sleeve. When we wore them, people would come up to us and say they were familiar with their work, which made it lovely.

Rambling Men can be all kinds of artists. Route 66 once had its own resident Rambling Man artist, Bob Waldmire. He was an interesting guy. He was an artist and a cartographer and I'm rather envious of him. He was a kind of hippy, and he travelled up and down Route 66 doing highly detailed sketches, paintings and maps of things he saw along the way. I went to see the converted bus in which he lived and travelled – and it was brilliant. I was consumed with envy. With its solar panels, shower, stove, sink and bed it's just the cosiest wee place. His bus is now in the Route 66 Hall of Fame in Pontiac, and all his favourite things are still inside. What a great way for an artist to live! I wish I could have met him.

The first piece of art I saw when I was a boy was a bleeding life-sized statue of Jesus. It scared the crap out of me. I don't believe in organised religion, but I don't believe in disorganised religion either. When you're an evangelist you can tell people God's talking to you and they'll send you the money. But go up to any psychiatric unit and tell them God's talking to you – they won't even let you home for your fucking pyjamas. I don't get it. Fuckwits coming to your door. I don't care. Fuck off. Fuckless people. People who've never been fucked. A fuck-free zone. Here's what you can do if you're having trouble with religious nutters coming to your door:

Knockety knockety knock . . . Don't open the door. You call out: 'Are you here to tell me about God?'

'Yes.'

'Fine. I'll be opening the door in five seconds. I'm naked except for a fireman's helmet. I have an erection. The choice is yours. Five, four, three, two . . .'

When you open the door, they'll be specks in the distance.

———

The Jesus statue haunted me my whole childhood, but eventually I discovered better art that didn't scare me – and I even came to appreciate some of the art that did. My sister Florence used to take me to the Kelvingrove Art Gallery and Museum on Sundays, where I was introduced to geniuses like Salvador Dalí, Vincent van Gogh, Paul Gauguin and Rembrandt, and by the time I was a folk singer I was starting to meet some of the brilliant Scottish artists like John Byrne and Sandy Goudie, both of whom became my friends.

I knew loads of Rambling Men who liked to draw. If they were travelling, they carried a sketchbook that could be rolled up. Some of these rambling artists were good and some weren't. Others I knew made things from silver paper. I had a pal who made horses from the silver paper inside a cigarette packet. He'd take the silver paper off and make a horse with his fingers. It was an extraordinary talent, creating this wee horse with the mane and tail flying upwards like it was galloping. He'd give it to someone, and they'd go, 'Did you do that? That's amazing.' I knew plenty of guys who did similar things, made something with string or did wee tricks like juggling glasses without spilling the water. It was a way to meet people. I never did anything like that. My thing was being funny. And playing music. But I did harbour a wish to draw myself . . . I just never thought I could.

Many years later, I met a guy when I was in Los Angeles. He was a film-maker and he wanted to make a film with me. I can't remember his name, but I didn't totally trust him. Not because he was untrustworthy but because he was very naive about how to get the money for that film. I thought I would maybe be wasting his and my time. But during meetings with him when we were having a laugh, he said, 'Do you draw?' I said: 'I can't draw.' He said, 'You can, you know.' He got a pen. It was a pen almost like a fountain pen and it came to a head like a ballpoint pen. He said, 'Look at a thing you want to draw and just draw it. Take your time. If you get it wrong, do it again. You'll be amazed how much it will please you.' And it did. I drew my Walkman, and it was really quite good. I was so pleased I'd drawn a thing I'd recognised.

So that was all rolling around in my head when I sat down to draw for the first time when I was on tour in Canada. I used to go out walking in freezing weather – and that's what started my art. I was in a hotel in Montreal, and I was bored. The TV offerings were particularly awful, so I thought I'd go out for a walk. They have a kind of rain in Canada that turns to ice instantly and sticks to you. It sticks to your coat, and it sticks to your neck, and it sticks to your face, and it's sore. I stuck it out as long as I could and then I ducked into a shop. It was a pet shop. I looked at the goldfish and the budgies and the dogs and cats and snakes and then I'd pretty much done it. I thought, 'I'd better leave.' I went out and I walked along a wee bit and then directly opposite the hotel there was an art shop. I went in, and looked about, and then I found some coloured felt-tip pens and some sketchbooks and I thought, 'I'll draw. I'll mess about.'

I started to draw tropical islands, but they weren't realistic tropical islands. There were red-and-white-striped trees and other things not found in nature. It was as though a bunch of ideas and information were just falling out of me onto the page, and I liked it. That was my start. When I got back home, I showed it to Pamela. I said, 'I know it's crap but do you think it's improving as it goes along? Tell me the truth.' And she said, 'It is. It's getting better.' So, I kept trying. Experimenting. I don't know why I did it. I had no intention of ever having art exhibitions or even letting anybody else see them. I thought I might let my friends see them. Maybe one day I might be brave enough to say, 'Look at that . . . I drew this, what do you think?'

I kept drawing and I found myself doing some odd things. I'd draw stripes as the background, and the sky was stripy too, but then I changed it and started making the human beings stripy and the sky something else. I have no idea why I made those decisions, they just seemed right at the time. And I went on and on. Then I started sending them to my management office. When my manager's wife was pregnant, I drew a pregnant woman with a baby inside that you could see, and she loved it. I sent more and more drawings to the office and the staff all had favourites and were fighting over them. Then my manager sent them to a gallery and the people there wanted to be involved – although at the time I didn't know what that meant. I thought, what if that meant doing a certain number of drawings each day, week or month . . . I wouldn't want to do that. But if these people are daft enough to take me on as an artist and believe in me, then okay . . . so here we are. I'm amazed at the way other artists talk to me. They are encouraging, say they like my stuff. They seem to like the fact I do it. They don't treat me like they think I'm a stuck-up entertainer who sells drawings because he's famous, and I'm very grateful for that. Maybe some think that, but they don't say it to me . . .

———

During my concert tours to different cities in the world I was always on the lookout for interesting paintings, sculpture, drawings and photographs. There's a lot of fantastic art in New Zealand – great people doing such original work. Roger Walker, the architect, for example, creates colourful architectural fantasies that are definitely box-free zones, colourful structures that would fit a futuristic toytown. And there was no shortage of weird art. In Kaikura, I found a dress made of plastic breasts at the wearable art museum, and once I even came across a huge picture of Elvis made entirely of toast – in a supermarket! I saw a lot of fabulous contemporary sculpture in New Plymouth: a very tall kinetic sculpture that resembles a testicle on a stick, and mounted Smarties moving around in the wind – big, orange, pointy things like giant beaks pointing straight up. I saw Braille artwork too. In Wellington, even the Houses of Parliament are interesting – built like a beehive – and in Eltham (the dairy capital of the area) they commissioned a guy to make their city water tank look like a big wheel of cheese with holes. They're not lacking in imagination down there. I also love the traditional greenstone axes made by Māori people. Beautiful things. And I occasionally come across publicly displayed poetry, like the poem down at the docks set into the wharf written by Bill Manhire, New Zealand's first Poet Laureate:

> *I LIVE AT THE EDGE*
> *OF THE UNIVERSE*
> *LIKE EVERYBODY ELSE*

———

Lovely. Auckland is a great city. Wellington gets all the 'capital' type of attention. But Auckland was the first city I landed in, and the first city in which I played, and it has a place in my heart. It has terrific art shops. I bought some nice pottery there.

———

In Australia, I sat astride the big 'lost purse' sculpture in Melbourne. Fantastic sculpture abounds in Melbourne – I love *Angel* by Deborah Halpern, commissioned to commemorate Australia's colonial bicentenary, Melbourne sculptor Geoffrey Bartlett's *Messenger* in the National Gallery of Victoria. Brilliant. And outdoors you can see sculptures like huge, colourful sections of pipe, with children sitting round them having picnics. I like some of the supersized creations too – like the *Angel of the North*, the biggest sculpture in Britain, designed by Antony Gormley that looks over Gateshead. I've come to believe that people should always live among sculpture and paintings, so I love seeing large murals on the sides of buildings – such as those in the Shankill Road in Belfast, with its fantastic, colourful loyalist art. And in 2009 the local government in Pontiac, Illinois invited 160 artists to travel to the town and paint murals on the walls of the town buildings on Main Street. I suppose they wanted to make it cheerier. Some of the murals look a bit like Victorian advertisements; some depict events and people from the town's history, and they really do add life to the place.

It's fun when public art attracts controversy and irreverence, as has happened with the monument I saw in the centre of Dublin officially titled *Anna Livia (The Spirit of the Liffey)* but widely known by names like 'The Floozy in the Jacuzzi', 'The Whore in the Sewer' and 'The Biddy in the Bidet'. Love that. And in Cardiff I saw sculptures made from road signs that were immensely good. Good old Cardiff. All this and Shirley Bassey too . . .

They do a lovely thing in Sydney. They set up 'sculpture walks' so you can wander along a set path and see an open-air exhibition of works by Australian artists. There's one that was set up in the city centre, in the grassy areas of the Domain and Royal Botanic Garden, and another I've seen in Bondi, called Sculpture by the

Sea. You walk two kilometres along a coastal route above the sparkling water and white sands of Bondi Beach and Tamarama on the North Shore of Sydney, seeing all kinds of sculptural art set on sand, set into rocks, or planted on the grass. It's spectacular. And that's a great advantage of having a reliable climate – you can trust the weather to behave for something like that. If they tried to do it in Scotland, the artists would be crying. 'Oh noooooh! My wee papier mâché piece has gone soggy AGAIN!'

It was in Chicago that I met the brilliant artist Preston Jackson, who created fantastic solid-bronze castings of people in his community. He told me he rubbed some people the wrong way, and I said: 'Yeah? Me too.' Preston's African American bronze people are stunning – wonderfully amusing characters in poses that capture aspects of Chicago's street life. Preston was funny and enlightening. He and I were laughing about Pat Boone's version of Little Richard's hit 'Tutti Frutti'. Boone, who was white and Christian, obviously didn't know the words had highly sexual connotations – bible in one hand, unwittingly singing about sex. And Preston showed me the 'Green Book' – a disturbing travel guide for people of colour travelling outside their neighbour-hoods, originally published in 1936. It contained life-saving information about where they could sleep, eat, find doctors – and, most importantly, lists of places they should steer well clear of once the sun went down. It's mind-boggling what those people had to deal with – and in some cases still do. I was talking with Natalie Cole in Los Angeles a few years ago. She told me that, in the fifties, her famous father Nat King Cole couldn't stay in the hotel in which he was performing and couldn't even walk in the front door. He had to enter via the kitchen.

———

I came across all kinds of crazies during my American travels, like the Toilet Museum. I was a bit underwhelmed though. Once you've seen twelve toilet seats you've kind of done it, but the owner had five hundred all displayed on his walls. A toilet seat saved me once. I was playing in a town in Australia where Tiny Tim had played the week before. They had unconventional guests, and I was one of them. There was a town nearby called Iron Knob and, on my day off, the crew and I went to see it. We were in the pub in Iron Knob and there was a crowd of loud-mouth yobs hanging out around the bar. It was the kind of old Australian pub that's a very male domain. In the past, some of those pubs even had a trough running around the bar so men could take a piss while finishing their next pint. I was wearing earrings, which attracted some unwanted attention. There was some kind of trophy behind the bar – it was a toilet seat with inscriptions and miniature trophies on it. I got up to get a round for my crew and a big loudmouth by the bar said: 'Does she know you took her earrings this morning?' I replied: 'Who took your picture out the frame?' That relaxed him. Made everyone laugh. 'That's my boy!' he said, 'What are you having?' And he got the round. It's good when a killer line comes to you in a timely fashion. Especially when it protects you from a smack in the mouth.

Near Amarillo, Texas I visited the Cadillac Ranch. Stanley Marsh 3 was an eccentric millionaire who created some iconic graffiti-covered pieces of art in the seventies that represent the American love of autos. I saw large slices of Cadillacs lined up in the desert and pitched at exactly the same angle as the Great Pyramid of Giza. The artist encouraged people to repaint the slices with new graffiti, so every now and again he cleans off everything to create a new canvas. I'd never graffitied before, but I went for it. He had already written my name on some of the pieces. I wrote 'Hello' with black spray paint.

After I left the Cadillac Ranch, I began to notice the Californian desert was all around me. It's the kind of place where hippies in the sixties and seventies used to 'turn on, tune in and drop out'. I used to hear about it in the sixties and I was a bit jealous of that kind of life. At Elmer's Bottle Tree Ranch I saw another interesting permanent exhibition – sculptures made out of collected bottles and all kinds of things: guns, toilet seats, hubcaps, old fans and rocking horses – hanging off trees and displayed in other novel ways. It was crazy, highly imaginative work.

I once went to a museum of barbed wire, but I couldn't decide if it was art or not. I never knew there were so many different types of barbed wire – all with unique knots and designs, and each with a special name. 'The Devil's Rope', for example, was invented by Joseph Glidden. Before barbed wire was invented, there was free-range free grazing all over the state, but then people started fencing off their land, especially if they had water. This caused a lot of fights and even deaths. I was fascinated by some types of barbed-wire splices – those jaggy bits that cut your trousers. I always wondered why they made a right-angled cut. In the museum they even had a display of World War One barbed wire. You could see that the British had more spaces in their trench wire. Obviously, they were a gentler race of people . . .

In Darwin, Australia I visited a wonderful artists' community, populated by weavers, designers and painters. I was crazy about their art, especially the bright designs painted on cars and corrugated iron. The place was essentially a cooperative venture, where all food caught or gathered was brought back to share. I was so impressed. We could learn a lot from them. There seems to have been an explosion of First Nations art in recent years. I met Jimmy Pike, a famous Walmatjarri artist whose work is hung in all the major museums and art galleries in Australia, and in several international galleries. He lived in the Great Sandy Desert as a child, then became a stockman. After getting into trouble he went to Fremantle Prison, where he learned to paint with materials that were non-traditional for him. It was a privilege to meet him and watch him paint.

Several Australian artists have had a profound effect on me. Ken Done is brilliant. And he's so accessible. He'll talk to you all day about his art. He doesn't shy away or say, 'I don't understand it,' like I do. I took a seaplane across Sydney Harbour to film an interview with him. We landed in the water opposite his beach house and Ken picked me up in a little rowing boat with Spot the dog.

Ken paints with intense blues – Sydney Harbour or morning glories. He taught me that it's not important to depict something realistically, it's more about painting the *feeling* of, say, walking by the harbour at night when it's still hot and the sky remains warm. He helped me to really see things, to use my eyes. To notice when a storm is coming. You get great storms in Sydney. Pink lightning. Now I get out of bed to watch. Ken had seen my show the night before I interviewed him. He said: 'I envy the instant feedback you get. If I do something that's any good, then hopefully the dog will bark . . .'

While I was hanging out with Ken, he gave me a tutorial on creating art. He said: 'I'd love to be able to draw as well as a five-year-old. That freshness. People say, "I can't draw." I think people *should* draw and paint. It's only that they have an expectation that it'll be real. It should just be the *feeling* of a cow or a flower or a person.' I wasn't drawing much then, but I had started. I said to him: 'I've done some drawings. Can I send them to you?' And he said: 'Sure.' But I didn't do it. It's something to do with the art belonging to you and being part of what you're in. I learned from him I should love my own stuff and not to question it, or question myself about what it is. It's itself. It doesn't need to come up with answers for you. I learned from Ken that I don't need to explain it because it just . . . is, and that gives joy. I'm really grateful to Ken for teaching me that. He gave me courage.

The Melbourne artist Asher Bilu is brilliant too. I visited him and his wife Luba, and their bird Billy Blackbird who lived in their garden. After I left, I heard the bird used to ask where I was. I often feel very close to people who have given their lives to art. Brett Whiteley, for example, was not only a genius but he was a joy to know. He embodied everything that's playful and Australian. I first met him in Sydney at the swimming pool on the roof of the Townhouse Hotel. Mark Knopfler introduced me. Mark said: 'Brett's an outstanding artist.' But at the time, I wasn't sure what he did. Well, they say that about a lot of people. But Brett was so wild and full of life and ideas and running and jumping around. At first, I knew I liked him, but I didn't know what to make of him. He was wearing casual clothes and a funny little hat perched on his wild curly blond hair. He looked a bit like Harpo Marx. He was a truly beautiful guy, with a great mind. He loved all the books he had read and frequently quoted them. I sent him a copy of my favourite book: *A Confederacy of Dunces*. He started coming round to see me at the Townhouse. He'd just show up in reception and they'd call up and say: 'Brett's here', and he'd take me for Japanese food. He loved sushi. Had a great lust for life.

Spontaneous. Pamela and I invited him to our rented house in Pearl Beach, and he was immediately jumping off the verandah into the swimming pool. That was risky. He used to go to the zoo just to see the iconic Tasmanian devil, which is a rather ferocious meat-eating marsupial. The creature would never come out of his lair, so Brett would throw coins at him to lure him out and get him to snarl. The reason was that Brett loved the red colour of his mouth.

Brett was a great laugh and used to make me roar. He came to my show, then bet me a hundred dollars that I couldn't do a completely different show the following night with not one repeated story or comment. I did it! Two and a half hours twice! But when he threatened to come for a third night, I had to tell him to fuck off.

Brett gave me a plate with a painting of a little bird. He saw himself as a bird. Twice in my life little birds have flown in and made a huge impression. The first was a jackdaw that landed on my head in Rothesay when I was a boy and said: 'Hello!' I nearly passed away. The second bird was called Brett Whiteley, but the little bugger flew away at age fifty-three. He was a charming, funny, gregarious man who enriched my life beyond recognition.

———

Some art is just way beyond me. Once I went to the University of Northumbria arts and design degree show, where there were extremely compelling installations, drawings and paintings. I can't pretend I knew what was going on, but I loved it. I was presenting a TV show in which we were featuring the art show, but it was a struggle to know what to say about it. I never want anyone to think I'm narrow-minded when it comes to art – or anything else for that matter. I'd hate people to say: 'Billy dislikes irregularity', because then someone else would be bound to say: 'Okay, All-Bran for him!'

13

HOW TO BUILD AN IGLOO

—

IN SCOTLAND THEY say: 'If you don't like the weather hang on for twenty minutes.' Then, when the sun comes out, they say: 'We'll pay for this!' I love how dramatic the weather can get in Scotland. With that kind of wind, you can practise standing at 45 degrees. If you put on a raincoat in the Highlands you can hang-glide. Go out in a cape and land in Oslo. That's where all the trees are anyway. Where do dogs piss there? Lie on their backs and piss in the air? People in Bergen are going: 'Fuck! The rain smells like piss.' If you go on holiday there, don't take a beach ball. You'll throw it to a child and it will end up in Denmark. There are people going shopping with their weans on bits of string.

I presume you know the history of the Scottish people in New Zealand? Two men – Hamish Dreich and Donald Dreary – took a big boatload of Presbyterians to the country. They all had those wee disapproving mouths that look like an arsehole into which someone's just slipped a slice of lemon. They landed in the north of New Zealand, and the sun was shining, and all the Māoris ran out and gave them a welcome . . . and we know how that goes. 'Aggghhh! Look at those people with their tongues out!

Dirty buggers – we know what that means! People with tattoos . . .! We must go further, further, further south!' They land at the top of the South Island. The place was wonderful: beautiful sun belting down, mountains and palm trees . . . life's fucking great! 'Nope!' They go even further south to Dunedin: 'Horizontal drizzle – excellent! Let's stop here! Settle down. We can whine here for fucking centuries!'

I don't normally care about the weather, but when you're a rambling entertainer you *have* to care about it in certain places. If you're in Calgary or Alberta, you'll hear a notice on the radio in the morning – they tell you the limit to the time you could spend outdoors with your face exposed. It might be five minutes or twenty minutes, or whatever the temperature and conditions dictate. After that you have to get indoors. So, you have to pay attention and do as you're told. In other places with milder climates, extreme weather is just a nuisance. For example, if you're somewhere where it rains too much you get fed up being wet and, although there's no real harm in it, your clothes are soaked all the time. You have to wear something that's waterproof and carry an umbrella, and the climate can even affect your health. And, if you're touring as I did as an entertainer, you don't want to catch some weird virus that's going around in a particular place.

Rule number one: don't let your wife pack your bags. She'll put in things that *she* thinks are good for you to take. You'll end up with four jackets when you only need one or two. You need one casual one – maybe a denim one – and a heavy waterproof one. Don't take things you don't need. Pack smartly. Put things inside your shoes. And just have one bag. A Rambling Man would never have more than one bag.

When it's too hot, get off the street. Get a fan and cool

yourself down. It's a bugger, heat. Nothing you can do about it. Don't wear man-made substances as underwear. Wear cotton. Don't wear nylon or any of its cousins because you'll feel horrible. It'll make you feel as if you have a bunch of bananas down there. And your feet will stink. Not from the underwear, though, from the socks; wear only cotton or wool socks. And before you get off a long-haul flight, change your socks. You'll be astonished at how much better you'll feel.

When I stepped off the plane in Coober Pedy in Australia, I initially thought I had stepped into the exhaust of the aircraft. But then I realised it was the wind that was hot. Not warm – boiling hot. I'd never been anywhere so hot. I got on the motorbike and rode across the airport on it, but I couldn't cool down, even in the slipstream. I had to ride around a lot while I was filming in Coober Pedy, and it was hell. But I still always wore my leathers.

Weather must always be considered when you're riding your bike. In cool climates, you need to wear layers to fend off the worst of the weather. And when the weather's okay you should still wear your leathers in case you fall off. Leather saves you from losing your tattoos as you scream across the gravelled roadway. And if you can, wear something smooth like a shiny shirt under your leathers so, if you fall off, it'll help your skin slide under the leather, and you don't get scraped. It's also very sensible to wear gloves. As I said before, it's inevitable that you will meet the road at some point, so it's best to prepare for it.

————

The worst weather I've ever had while touring was about thirty years ago in Canada when it was a white-out for about three weeks. I really got fed up with it. Canadians are good at handling that kind of weather, but I got really tired of the constant blizzards. I remember being in Toronto and being driven down a street in the financial district around midnight. I'd done a gig, and we'd stayed behind at the venue for a bit at the end. I was on my way back to the hotel. When we stopped at the traffic lights an immensely powerful gust of wind came off the lake, hit the back of the car and shoved us forward about six feet. It was extraordinary to feel the weather messing you about like that. It was serious. You can't walk about in weather like that, it's not pleasant.

I don't really mind being cold. It's a natural state for me. But there has to be a point to it. Travelling in the Yukon had a great point to it; it lived up to my fantasies. There's still some of the architecture left from the old miners. The buildings are often leaning to one side – they're built on frozen matter, so with the years passing and global warming they have begun to fall. There is still a hillbilly atmosphere in the pubs. In one the locals test you by giving you a drink with a thumb floating in it and say you must let it touch your lips to be considered a real guy; it's part of some local legend. Some of those male-dominated traditions make me uncomfortable. Like men's toilets where men pee in a trough – it's fucking primitive, isn't it? Women get a door and a wee room each – men get a trough. Men sometimes come up to me while I'm peeing.

'Are you Billy Connolly?'

'Yeah . . .'

'Oh hello!' Hand out.

'Fuck off!'

The coldest I've ever been was when I was in the Arctic. I travelled forty-eight hours from London to the Canadian

Northwest Territories, ending up at the most northerly airport in the world. Then I drove a skidoo further north. I had a brilliant trailer for my skidoo on which I loaded my rucksack, roll mat, cooking stuff, sleeping bag, groundsheet and extra clothes. I carried a saw to cut the ice, an axe, a knife and a Lee and Enfield rifle – the same kind I'd had in the Paras. And, of course, my banjo – on which I played 'Campbell's Farewell'. I didn't sing it, though. All that crap about Bonnie Prince Charlie would hardly have fitted the occasion:

The prince who should our king ha'e been
He's wore the royal red and green
A bonnier lad was never seen
Than our brave royal Cherlie

———

With my trusty trailer piled high with my necessities, I drove to a town that was home to 135 Inuit people. They were delightful. They built me an igloo, and I watched very carefully because I was told that, later on, I would have to try to make one myself. Next day, after travelling even further north, I found a place to make my camp beside an ancient iceberg. This was when I was going to have to try to make my own igloo. There's something very basic about making yourself a shelter. Every time I've seen a house I wanted to buy, I'd be standing in the grounds of it and would have a desperate desire to pee. It happened to me again beside that iceberg when I was choosing the site for my igloo. I was bursting to pee, so I must have picked the right site. I suppose it's a primal need to mark the territory.

I tried very hard to make my own igloo, but I made a mistake early on. You have to build the ice bricks very carefully, laying them at exactly the right angle so the walls will be strong. If you make a mistake early in the process you're doomed, because it's all about the angles. The bricks of snow are slanted and must be placed accurately so it all goes up like a cone to a narrow bit at the top and then it closes. That's what I was taught. It's all very well knowing that, but when you're freezing, and you have to dig all that bloody snow out of the ground, it's not easy. The guy who built one for me had been doing it all his life. He was about seventy-odd and he built the Taj Mahal of igloos before my very eyes. It was a thing of beauty.

Igloos are lovely inside. When you get inside through your little tunnel it's wind-free. You can hear the wind outside and you feel so clever. You light your lamps, put your groundsheets down and the room begins to sweat and seals itself up. It's delightful. A very comfortable way to live. But when I tried to build an igloo, it went wrong about a third of the way up and started to collapse on me – so, I quit. I remember sitting on it talking to the camera. I called it a pyramid by mistake.

I wish I had done it better, but I was very happy to have made a third of an igloo. I ended up in a tent pitched in the ice. To pitch a tent in that kind of environment you first dig a hole with a spike. Then you dig another hole nearby to make a little tunnel you can put your finger through. Then you can attach the guy rope and tie it to the ice. Lovely. It was a great tent. Unfortunately, I started to have a recurring day-mare that the ice was going to crack and I was going to fall through in my sleeping bag and drown. That was stupid of me, because six-foot-thick ice doesn't crack so much that you could fall through. But once I calmed myself down, I found the tent cosy. There was a musk-ox skin on the floor of my tent and a good sleeping bag – in fact, I was so cosy in there I didn't want to come out. I wanted to stay in and play my banjo. I had yak-skin ground-sheets and sealskin clothes that smelled to high heaven, but nothing bothered me. My water supply was an iceberg – pale blue in colour and consisting of frozen fresh water. It was the most glamorous water tank I've ever had. I was very happy. And there was nobody north of me in the entire world.

When I went on excursions, I travelled on dogsleds. The dogs love their keepers, but they're out of control. They're not pets. They are wild and hardy, and they sleep outside in the snow. People throw them frozen meat – whale and seal – and they have to lick it to make it soft before they can eat it. That type of food has an extraordinary effect – they fart like there's no tomorrow. You're out there on the virgin snow, just the dogs and you swishing along, and it's wonderful – except for the clouds of fart floating back at you. The air goes blue. And when they shit, it sticks to the hair on their bums, so when they fart the shit comes flying off towards you. You're just enjoying your-self in the clean air when suddenly this overpowering wave of awfulness hits you.

———

Overall, it's good fun. It's great being the guy in the Arctic. I ate commando food. SAS supplies are not the jawbreaker biscuits that military food used to be. It's brilliant. It comes frozen, in an envelope. You add water and heat it up, and it's delicious. There are colossally good stews, and desserts such as apple pie and custard. Classy. Right up my street. And I was given a nice little cooker. I wore caribou-fur socks with sealskin over the top. I just put my bare feet inside them as I was instructed to do, and my toes were warm as toast. Okay, I wouldn't have been very welcome outside Harrods in my politically incorrect clothing, but I didn't give a toss.

I was alone for long periods and really liked it. There's a silence in the Arctic you'll never experience anywhere else in the world. It's a silence you can hear. You actually notice it. You say to people, 'God, it's quiet, isn't it?' As if the silence just made a noise. It has a weight, a presence.

For a while, it was sunny most of the time during the day, and then the snow came, and it snowed for about three days. It was lovely fresh snow to ski around in. It doesn't rain in that area. Well, it rains little twinkly rain. You know when you see a snowflake magnified and it has those little legs and a geometric shape? Well, millions of little ones come down and land on you. You're just waiting for some music to start going, '*tinkle tinkle tink*'. I thought they should call it fairy dust – I thought my girls would really like it. When I was on the road I always made messages for my girls. I would cut out pictures from magazines or books and paste them onto a sheet of paper with wee tidbits about whatever place I was in then fax them home. But in the Arctic there was no way to communicate. I was completely isolated from the outside world – and I liked that more and more as the days went on.

I began to get annoyed that the camera crew came round during the day, filming my every move. I wanted to have more time completely alone. I became very possessive about the virgin snow around my tent. The crew all arrived in skidoos and trampled all over the place. I knew I had no right to be like that – it was selfish and mean – but I couldn't help it. They would throw cigarettes in the snow around me, and I would say, 'Look at the state of this place! Not good enough!' I began to complain. 'You've got to treat it with respect. It's where I live!' I was being sincere about it. It got to me because the film crew would all go away at about five o'clock in the afternoon and then it was just me alone in the Arctic.

When I say 'alone', there was a soldier from the SAS – Paddy from the Irish Guards – camped away at the ocean, watching for polar bears. He had said that if they were coming towards me, that's where they'd be coming from. I never saw any around there, but he used to tease me. He'd get on my two-way radio and go, *'De doo de doo de diddily doo'*, singing the 'Teddy Bears' Picnic'. It was good fun. But I had to be ready for polar bears coming in case they sneaked past him. There's only one way to tell if a polar bear's there, which is if it growls. It's unlikely to do that, though. If it fancies you, you're done. I kept my gun loaded. I was told that if it's far away, you fire over the bear's head to scare it off. 'What's "far away" in polar bear terms?' I asked. 'More than three hundred yards.' 'But what if it doesn't turn tail?' I asked. 'If it's still coming at you, that's the time to kill it for your own safety.'

Truthfully, I think the Arctic type of conditions provide the ultimate challenge for a Rambling Man. On the one hand it's the very best environment for someone who wants to wander alone and be self-sufficient, but on the other hand you are reminded that there's no salvation if you get into real trouble because there's no bugger anywhere nearby to give you a hand. There were psychological challenges too, the type that most Rambling Men do not encounter. For example, there was an iceberg outside my tent. It was considered a wee one, but it was about twenty feet high and fifteen feet wide. And that was just the bit you could see. Icebergs are made of fresh water floating in seawater. The seawater goes up and down with the tide, shoving the iceberg up and down so it rubs against the sides of the hole it's in and makes a noise like a polar bear. *Grrrrrrrooooogh.* I'd hear this when I was inside my tent and go, 'Oh fuck!' I'd get out of my sleeping bag, get my rifle, put my jacket on. I'd go outside – nothing there. Then I'd take a second look because somebody told me that polar bears know that their noses are black, so they cover them with their paws. That's a wonderful thing, but I had to be doubly careful. It was always daytime while I was there, so I just went to bed when I was tired. Eight hours later I'd get up and check my traps, then wander around looking for footprints. I felt like Grizzly Adams. I saw prints of foxes, seals and birds. In the evening the aurora borealis would come. I'd stand in the snow with this extraordinary exhibition just for me.

I developed a siege mentality – saving bits of string and so on. I'd tell the crew to bugger off to their hotel and leave me alone. I wasn't afraid – I actually became exhilarated. I'd get up, play a bit of T. Rex on my banjo, go for a wee run a couple of hundred yards, then come scootering back and make myself some porridge. Not a care in the world – and a wee bit of T. Rex to cheer me up. I filmed daily video diaries by myself. During one of them, I said to the camera: 'You're sitting at home, eating your fish suppers by your electric fires. You're wearing your casual clothes and slippers . . . and you're thinking: "Where does he go to the bathroom?" Well, I'm not telling you.' In fact, I dug a hole in an area not a million miles away and had to bare my arse there. It's not easy. You have to be quick. That was one of my main worries ahead of my journey. I read the SAS survival hand-book: 'In severely low temperatures it's best to do it in the tent and use the heat.' I thought: 'Not this soldier. I'll just be a wee bit colder and risk frostbite to my bum as I go somewhere else.'

It was a bit like the wilds of northern Scotland, where the environment is always bigger than you. It forces a kind of humility on you, where nature is saying: 'Watch your step, sonny boy. One step out of line and you'll be in terrible trouble here.' There wasn't a single living thing there that I could see. Usually, you see a wee seagull or something – not there. Absolutely nothing. The iceberg creaked and croaked all night, every night. They are millions of years old. Straight from the glacier, and well slippery. But after a while, the noises of the iceberg didn't frighten me so much. They kept me company. My Inuit teacher David said that there were people in them. His ancestors were encased in the ice. You could feel a presence there. The silence is so extraordinary in its density that you can feel it. There was another iceberg nearby that looked like Jesus – a man with a beard with a crown of thorns type of thing – but I was the only one who could make that out. Maybe I was losing it . . . I didn't care a jot if I was. In any case, I was damn sure that possessing a Rambling Man sensibility was the best way to survive that kind of test.

The Inuit were delightful people who seemed perfectly comfortable in that harsh environment. They had nice ways and customs. If they have a baby, they don't name the child till they see who he or she is. Until they perceive who has come back among them. They'll spot a trait in a baby of somebody who's died, and they decide it's that person. So sometimes a boy will have a girl's name and there's no problem with that. The names are for all. It's a lovely thought to have your mother's name. And Akiatuk, my guide, told us he had been given his mother's father's name, so he would give his mother a bad time if she wasn't behaving. He'd say, 'Daughter, behave.' And when they hunt and catch a seal, they don't get to eat it – it's for everyone else in the community. And when others go hunting, they'll get some.

———

Towards the end of my time in the Arctic I moved to the other side of the bay, about six hours away. I crossed the sea and came upon Craig Harbour, but there was no harbour and no Craig. Just ice. There was a post office with bear-claw marks where a furry creature had tried to get in the window. A huge, hairy, hungry thing. Despite that, in my tent that night I felt I was in the most serene place in the world. And I learned I had a ringing in my ears. I hadn't washed for five days, and I didn't want to wash. There was a blizzard on the fifth night, so I had to use a polythene bag to pee. It was frozen in the morning, so I went out and buried it with great ceremony.

Later that day I came across some polar bear prints around a bloody seal carcass that it had ripped apart and scoffed. Foxes and ravens had had their go too. It was very sobering. I saw a rather beautiful seal looking at us, then it nervously dipped into its hole. Apart from that, I didn't see any wildlife. But the creatures there are all white, so they are very hard to see. There could have been millions of them around, but I'd never have known. I set out to try to make it to an ice floe, but it suddenly turned very cold. A white-out. Formidable and dangerous – I had no idea where I was. It was a real salutary lesson about my puniness. I was just a city guy adrift in a solitary place. Despite my rambling ways, going to the Arctic allowed me to understand that Rambling Men are drawn to places where community thrives in some way. It doesn't need to be densely populated, it could just be a small group of people, but to be alone in the wilderness all the time would be difficult. I made it back to my base, knowing I was leaving the next day, and felt relieved I hadn't been eaten by anything . . . but there was still a whole night to go.

——

I learned a lot in the Arctic. I loved the solitude, the ruthless wildness of it. And I found that I could be comfortable on my own with just my thoughts – at least for a while. I'm not nearly as afraid of myself and my imagination as I used to be. And I'm not afraid of extreme weather now, because I've been in the midst of it – high, low, wet, dry, and everything in between. Mind you, I've seen some devastation from winds. Three hundred miles outside Chicago I was on my trike and got caught in a rainstorm at the beginning of tornado country. The county of St Louis has more than its fair share of these natural weapons of mass destruction – about three hundred per year, and there were five of them just before I arrived there. I drove into the remains of a township that had been completely levelled by a tornado that was 166 miles long and one and a half miles wide. It had taken just thirty seconds to destroy the entire town. A man who survived it said the instant his ears popped he knew it was coming. 'It sounded like a freight train,' he said. A family was searching for valuables in the rubble. One woman had managed to find her engagement ring. Her wedding dress had survived too, just it was now covered in mud with a bit of a hole in the seam. But those people were extraordinary. They'd lost everything, but their mood was high. They were already redesigning their homes with an extra bedroom and a bigger porch. That's pioneer spirit for you. Maybe we should stop complaining about the weather. Try harder to count our blessings.

It might seem churlish to say, 'Things could be worse,' but did you know that tornados and waterspouts have been known to sweep alligators into the air? Imagine one of those big bitey reptiles landing, alive and kicking, right next to you and your pancaked house.

I've never been in a hurricane. As a matter of fact, I'm secretly troubled by the fact that Pamela and I now live in a place that's regularly battered by winds that could whoosh you off without so much as a 'please' or 'thank you' and dump you in the jungles of Cuba. We had a big hurricane just after we moved here. As soon as Pamela heard it was coming, she forced me onto a plane to California, but I've told her I want to stick around for the next one. The authorities order everyone to leave but the locals stay and have wild parties.

I went to a hurricane party once, but the hurricane never showed up. It was when I was touring with Elton John. We'd been playing Madison Square Garden. After we came off, word came backstage that there was a hurricane warning for Long Island, and that someone was holding a hurricane party. We all got into buses and cars and set off to join in the merriment . . . but the guest of honour never appeared. It was fun, though, being in a posse of celebrity storm-chasers. We were frankly quite ill-prepared to meet a hurricane, but we weren't bothered; drugs and alcohol were probably involved . . . yeah, drugs and alcohol were *always* involved.

14

THE LAST RESORT

—

—

SOME OF THE nicest people I know are decomposing as we speak.

I've always liked graveyards. I like reading the headstones. Lots of them have Bob Dylan's line 'Forever Young' written on them. Pish. We're not forever young. We're forever decomposing. 'Forever Dead' would be more fitting. There's a lot of hogwash: 'Asleep'. I don't think so. Dead, methinks. And some are funny and savage. 'Stick your nose here and I'll set about you'. I was thinking I'd like: 'Jesus Christ, is that the time already?' on mine, but Pamela was shaky about it, so we settled on 'You're standing on my balls!' in tiny wee writing.

Glasgow has some rather nice cemeteries. A fan once called out to me in the street with an accusation: 'Eh, Big Yin? You're just like us . . .!' 'What d'you mean?' 'You drank wine in the graveyard, didn't ye?' Now, that was a bald-faced lie. I have never drunk wine in the graveyard in my life. But, okay, I've shagged there . . .

—

My favourite graveyard in the world is in Egypt. It's the ancient site called the City of the Dead. It's a remarkable place, about four miles long, where all the dead people have houses. There are thousands of them. It's just like a massive, eerie housing estate. Years ago, I did a concert in Cairo for a Scottish guy, who asked if I'd entertain Scottish workers who were restoring a famous old grand hotel. He provided an expert guide to show me round the place. She translated many of the inscriptions on the tombs, and I thought it was wonderful. We mainly hear about pyramids in Egypt, and that's what you got if you were a real head honcho in ancient times; if you were a pharaoh, you got the full gold pyramid with all your stuff inside. But in later centuries the people with serious money found their last rest at the City of the Dead. Bereaved families used to stay in the mausoleums for the customary forty days of mourning and would revisit the deceased person for memorials and vacations, but essentially the buildings remained empty – until more recently, when living people started squatting in them. Nowadays, many thousands of breathing 'tomb-dwellers' live there. That's Cairo's answer to poverty and the housing shortage.

The Necropolis in Glasgow is another beauty. It's a Victorian cemetery that's the resting place for many of the old Glasgow worthies, like the nineteenth-century author William Miller, who wrote 'Wee Willie Winkie'. His gravestone says: '*The Laureate of the Nursery*'. The Necropolis is where people with serious money found their last rest. The tombs and headstones and monuments were installed at vast expense. These people had bigger houses when they were dead than most people in Glasgow had when they were alive. Rich merchants are there, and many people who created and maintained the second city of the empire, as Glasgow was known. Charles Tennant is buried there – his main business was a chemical works for processing coal at St Rollox. Sounds like rhyming slang to me. Some of the Lords Provost are buried in the Necropolis, as well as shipbuilders, artists, publishers and engineers and war graves from World War Two. And there are some lovely tombstones and monuments. It's a picture of how Glasgow used to be.

I've never been scared of graveyards. In fact, I always felt kind of welcome there. I used to walk around in graveyards when I was a kid, but I haven't the foggiest idea why. It just felt lovely in there. I wanted to walk alone there and be at peace. It was a bad time in my life. I couldn't get on with my teachers. I couldn't get on with the adults at home – my father and his two sisters Mona and Margaret. They all found me a dead loss. As far as they were concerned, I was a complete waste of space. Once I volunteered to go to the funeral of a bishop that was to be held at Glasgow Cathedral. I had known him because he had been a priest at a parish near my school. There were all sorts of priests there, from the lowliest to the highest. I went with two other guys. But I had a hole in my jacket, and the teacher in charge of us said, 'Connolly! Why are you wearing that jacket?' I said, 'It's the only jacket I've got.' He said, 'You were warned about this before you volunteered – that your uniform had to be in very good shape.' And he sent me home. That was all because of my Aunt Mona. She would never repair anything. If something got a tear in it, she'd make it bigger. Once she set my school tie on fire. I wasn't there at the time, but I found the remains. She was a case.

———

I did a whole TV series in the USA about death called *The Big Send Off*. It felt good talking about death and telling the truth about it. You can only lie about it or avoid it for so long. As you learn about the Buddhist way, the Mexican way, the Islamic way of dealing with death, and all these different ways of dying and being buried – you can't kid any more. You have to view death from a practical point of view. There's a great sense of relief about doing that – a feeling of release. Nobody was expecting anything of me when I made the series – just my thoughts – so I found myself being very honest about it. I explored the one certainty in life – that we're going to die. You see, reports of my recent demise have been greatly exaggerated. There was a week a few years ago where on Monday I got hearing aids, Tuesday I got pills for heartburn, and Wednesday I received news that I had prostate cancer and Parkinson's disease. But despite all that, I never ever felt close to dying.

I'm fascinated by death and how we deal with it. We skirt around the subject of death but it's a 21-billion-dollar-a-year industry in the USA. I visited a pet cemetery where a burial for a guinea pig cost $550 for the plot, $350 for the coffin and $1k for the marble headstone. Then I went to a funeral directors' convention in Texas, where they were promoting embalming fluid party packs, shampoo for dry, lifeless hair, blankets with life-size pictures of the deceased – you can sit with it for a drink or take it to bed with you. They had zombie-proof steel coffins on display, hot-rod hearses, and you could reserve rockets to the moon for your ashes. There was the eco death-suit threaded with mushroom spores; the fungi feed on your decomposing body and take you back to the Earth as human compost. You're of no use to any Earth-dwelling person, so it gets rid of your body in a sensible way without bothering anybody. Not sure how I feel about a mushroom eating me. There's a theory that mushrooms come from space. They came on meteors and stayed. I believe it. They breed unlike any other species on Earth. And they look like aliens, not like anything from here. So do octopuses.

There are many choices now when it comes to how you can be buried. The Neptune Society disperses ashes under the Golden Gate Bridge, and no relatives attend, because that's the way the deceased person wanted it. I went along, and thought it was profoundly moving, with the ashes trailing in the water and a wake of flowers following the boat. I'd rather like that for myself. My pal Eric Idle said he wants fireworks and an element of dressing up. Maybe even a cashpoint in his cask so it's a useful visit. His song 'Always Look on the Bright Side of Life' has long been a favourite funeral song.

Some say you only die the last time somebody says your name. The Mexican Day of the Dead is about remembering those who have passed. Some people get memorial tattoos. One woman got a peacock tattoo for her granny because she always had a bunch of peacocks wandering round her property. It was only after she got the tattoo that she found out her granny hated the peacocks and used to shoot them.

I met some people who think science is the answer to death and can offer eternal life. There was a whole team of scientists who believe ageing is a disease and therefore curable. In Buddhism, there's a meditation that involves when you see a dead thing on the road you say: 'That is the Way of all things, and it will be the Way of me too.' I like that. I met a guy who said he believed in everything – 'That way I can't be wrong!' At the time he was ill and about to die. He let me try his coffin for a dollar. It was like a box you get clothes in from fancy shops. It was made of cardboard, all ready to go into the flames. According to his wishes, he was buried in a skeleton suit. Before he was cremated, they took the skeleton suit off and sent it to me. Pamela and I were living in a tenth-floor apartment on Fifth Avenue in New York at the time, and this THING arrived. It was actually a great skeleton suit made of rubber with zips, but I don't dress up at Halloween and I didn't know what to do with it. I didn't want it in my life, so I had to think about a fitting way to dispose of it. In our apartment building they had a chute that you put your trash into to send it down to the bins on the ground floor, so I decided to chuck the suit in there. Sent it down to hell.

As for me – I haven't made up my mind about my burial place, but I'm thinking that instead of a headstone, a table on an island in Loch Lomond for fishermen to picnic on would be nice. During *The Big Send Off* I especially liked visiting the 'green' graveyard, a nine-acre plot of land in Texas, called Eloise Woods. It was opened by a woman called Ellen Macdonald, who was a neuroscientist. It was a very simple concept – you could be buried in a very organic way. Someone would dig a hole, drop you in it, and invite your friends to say 'Cheerio'. If you want, you can get buried with your pet . . . although he might not like it if he's not dead. It's charming. Clean. Good. Everything about it is good. One guy was buried in his chair. They just sat him in his chair and covered him with earth. Five million gallons of embalming fluid are poured into the ground every year, which isn't doing the planet any good, so this is the answer – a green burial ground. Families can dig the grave themselves and can mark the grave with something simple like a flat stone. It seemed a refreshing change from the denial of death in our culture where you put make-up on people to look like they are alive and simply sleeping.

I also liked the Columbarium I saw in San Francisco, a huge

structure that is a repository for human ashes. There are walls of remembrances where people can leave things the deceased person liked, displayed as dioramas – little plastic soldiers, flowers – things that meant something to both of them – on a little shelf. There were lots of snow globes – from Hawaii, for example, to remember a great holiday. And pictures of Scottish guys in kilts who'd died a long way from home. People do lovely things when they're left to their own devices. Some people are so imaginative and playful – even when it comes to their own death. The best headstone I ever saw was a hat and beer-keg. This man had his own coffin made while he was alive – out of corrugated iron. He used it as a wine-rack in his living room while he was alive, and then he was buried in it.

Making that documentary changed me. I used to think about death, and about the life I led, and would ask myself, 'How will I be held responsible for it when I come to judgment before God?' I don't believe that any more, although the whole thing is still a mystery to me. It was very comforting meeting those people I spoke to during the series. They were all lovely; there wasn't a single nasty person I encountered. The drive-through funeral parlour in Compton, Los Angeles, was a gas. Mourners could sign the visitors' book outside, then drive through to pay their respects. There was a dead woman in a coffin in the window. They sat the coffin up a bit so you could see her better. Peggy Scott Adams was the proprietor of the parlour. She said: 'Death is a part of life'. She was great. She made her clients feel good. She'd been a backing singer in a soul band. When we went through to the chapel, we had a little dance and sang, *'Goin' to the chapel and we're gonna get ma-aa-arried . . .'*

Hillside Memorial Park is LA's premier Jewish cemetery – with residents like the Max Factor family and Al Jolson. It's a bit like the theatre – being interred at eye level commands a premium, while the cheapest seats are up high, 'in the gods'. And you don't have to be stuck with it if your circumstances change; there are cemetery brokers who can help you cash in if needs be. Sometimes families will disinter a body, have the remains cremated and sell the plot. I learned there are even theatrical agents for dead people. I saw a lovely inscription on one headstone:

'In memory of Sophie Kravitz. I met my wife at a travel agency. She was looking for a vacation and I was her last resort.'

By contrast, just outside Los Angeles a lovely Muslim holy man called Gulad walked me through the way Muslim people are buried. He talked about simplicity in death. In Islam the family has control over the washing of their loved one's body. Death is seen as an equaliser. People are taken to the graveside in a simple van. If people can't afford it, the mosque pays. Gulad was one of the nicest people I've ever met. He was in charge of a mosque and he was so nice to the dead and so nice to me. He said: 'We wash our own people. And then we take them in a van to the burial ground.' I was laughing about the van, and so was he. I said, 'I think I'm a van guy.' He said, 'Oh, I'm definitely a van guy.' In his traditional way of death, the loved ones take the deceased person and climb into the grave with them and point them to Mecca. They put the earth around them then cover them. Gulad talked about it with such joy and understanding. I was deeply impressed by him. He took me to the mosque and showed me where to kneel. The carpet is specially designed to show where your knees should go.

He said: 'Do you believe in God?'

I said: 'I don't know. I have had trouble all my life, thinking "Does he believe in me?"'

He put his head back and roared. He said: 'Can I use that?' I said, 'Yeah.'

We got on like a house on fire. It was wonderful. When we die everything is so different. If our relative dies he's taken away in a box and we don't get near him until they bring him back in a fancy polished box and he's wearing make-up. Our style of burial seems so ridiculous beside theirs.

———

My daughter Amy was with me while I was making *The Big Send-Off*. She was the researcher, because she had studied Mortuary and Forensic Science. She had all kinds of stories. At the gravesides they told her to watch the mourners for any 'jumpers'. Apparently, people frequently leap into graves and get injured. Break their legs. Amy was taught how to identify, stop or catch them. It's weird, the feelings that come over you at funerals. The father of a friend of mine was lowered into the grave and I found myself crying. I'd never set eyes on his father before – I was just carried away with the emotions of everybody else. Then I remembered a line from a Gerard Manley Hopkins poem, 'Spring and Fall', we read at school – and realised it must be about my own mortality:

It is Margaret you mourn for.

———

But no matter how nice your funeral is, you can never predict what will happen in the future. John Knox – leader of the Reformation in Scotland – insisted he wanted to be buried within twenty feet of St Giles' Cathedral. At the time there was a grave-yard adjacent to the kirk, so it was a reasonable idea. But the area was eventually paved over and is now a car park. So, Knox's grave ended up being marked by a wee plaque at parking space number 23. It just goes to show you: all those people paying a fortune for their burial sites – who says they're really going to end up there? There are all kinds of wee tricks that council members can play if the price is right. Zoning? Pish bah pooh.

During the making of that TV series, I didn't leave my heart in San Francisco – I left another part of me in New Orleans. I had terrible pains in my stomach region, and they discovered that two bits of metal from a previous operation had come adrift. They had to pull them through my willy. I know you just winced. Surely that's a fate worse than death?

The best death scene I ever witnessed was in a pub in the Byres Road in Glasgow. I was having a pint and waiting to see if my uncle turned up. He didn't, so I just decided to have my pint and be on my way. But a guy came in looking kind of flustered, and he did that Glasgow thing I love – he blurted out something in the midst of our conversation. 'Och, it's fucking murder!' he cried. We were all sitting round looking at each other, mystified. He continued: 'I cannae go in . . . I dunno why I went in the first place!' I said: 'What's that?' He said: 'I'm at a wake upstairs.' Turned out there was a tenement above the pub, and he was in the top flat. He said: 'I don't even know those people. It's my wife who knows them.' He said: 'They're being nice to me, but I don't know them. It's a weird atmosphere. I wish I hadnae come.' He must have sensed my attempt at being sympathetic because he turned to me and said: 'I have to go back up. Don't suppose you'd come up with me?' I said: 'I don't know them either . . .' He said: 'That doesnae matter.' He said: 'We're friends.' We'd only just met but that didn't seem to be an issue. He said: 'We'll get a half bottle of whisky . . .' And we did, and away we went upstairs. See, this is an established ritual in Scotland. You go to the house where the dead person lived and sit around in the hallway having cups of tea and coffee and talking pleasantly about them. And then the official thing starts – someone invites you to come into an inner room where the coffin is to say goodbye. That's where you're given a drink and you join in the official toasts. You wish them all the best, and people say how much they liked them: 'He was a good man. An honest man.' And a funeral was always a good excuse for the local alcoholics to get a few jars in them. They'd just join the line. Nobody liked to chase them away. They'd put on a black tie and pretend to be a friend of the dead guy. Everybody knew – it was the same guys every time.

So, there we were, sitting around drinking whisky and making faces at the other people. We were trying to make conversation, but they were a rather stuck-up crowd – and I never did find out how the man's wife knew the deceased person. But then the door opened, and an extraordinary-looking head came round. This head had a huge Adam's apple, gigantic cheekbones, a great bulbous nose and pointed eyebrows. He was a desperate-looking man. And he said: 'Hello! I'm the corpse's brother.' Those were his exact words – one of the most hilarious and bizarre lines I've ever heard. 'Hello! I'm the corpse's brother. Would you like to come in the room and drink his health?' Well, I just had to go in. And anyway, there was a nice atmosphere and whisky to drink his health – to see him on his way. The deceased man was lying in an open coffin, and I couldn't help noticing that he was remarkably similar in appearance to his brother.

———

I've always been fascinated by the rituals around death. There's a lovely thing that's happened in the Clyde. Govan Road in Govan, where the shipyards were, runs exactly parallel to the flow of the River Clyde. And when someone in the shipyard died – someone known and liked, like a shop steward or a foreman or a manager – they would do a parade along the road. People would line the pavement, and they'd take off their caps as they came along in a kind of wave, like those dancers in New York – the Rockettes – but with their bonnets. It was lovely, seeing the wave of caps being doffed all along the street. It was a nice sign of respect. I learned that when American hobo communities hold funerals, there's a similar show of respect. Everyone wears a strip of burlap, and they tap the gravestones with their walking sticks. The hobo euphemism for 'died' is: 'caught the Westbound'.

I like funerals too. My manager used to be mystified. He'd say: 'Billy – even if you haven't seen someone for twenty-five years and they die in another country you'll jump on a plane.' It's true. I just like to see people off. It's important to me. Boys from school. Apprentices from the Clyde. Welders. It's a bit like a school reunion. Seeing all the people from that place and time – that world I was once part of. I spoke at Jimmy Reed's funeral. It was lovely. I just told funny stories about him. About how he sang. '*Passing strangers now . . .*' Crooning. And the people were roaring with laughter cos they remembered him. But I'm not the only one who likes funerals. In Ghana the people love a funeral so much that there's a demand for a four-day week so people can get over the hangover. And what's not to love about some Taiwanese funerals, where the family hires 'electric flower cars' – trucks with mobile stages for strippers?

We all know you have to watch how you speak about the dead. When my friend Stan's wife died, I wrote to him and said: 'She always used to sneak up to my end of the dinner table to have a good laugh. I always remember her like that . . . having a laugh.' Stan wrote back: 'Yes, I remember it well, Billy. She often spoke about how bored she was . . .'

EPILOGUE

COMING HOME

AFTER I'VE BEEN on the road for a while, I start to become a bit nostalgic about home, especially if my family sends me adorable pictures of them and my wee doggies. But once I do arrive home, I find it difficult to be at peace. My mind is still on the road, and my body is still in travelling mode. I get used to hotel living, so much so that Pamela has often complained that I have 're-entry problems', that I arrive home and try to dial room service. It's not easy to go from being Sir Billy Connolly, lauded comedian and artist, with everything laid on by my promoters for my comfort during a tour, to being a dad who forgot a birthday, or a husband who failed to take out the trash.

But once I settle down, I enjoy a simple routine. Mine is: wake up, shoo the dog off my bed, switch on the TV and try to find football, have breakfast in bed, shower, write, draw, have lunch, do my fucking physio exercises, go for a walk, play my harmonica, write or draw some more, watch the news on TV, have dinner, watch a TV show about grisly murders, ghosts or monsters of the deep, sleep . . . and repeat. I like going to the movies – in an actual cinema – and sometimes I go to the beach, or out in a boat to fish . . . and, of course, there are doctors' appointments. But basically, that's it. I realise some people might find that a tad exotic for their taste, but to me it's a wonderful life.

If somebody said: 'Billy, I'm going to take you on one final ramble wherever you want to go. You can have as many stop-offs as you like or take as much time as you want. You don't have to be filmed, you can just have fun and enjoy it in complete comfort . . .' where would I want to go? Well, I'd like to go to Australia and have a nice few weeks swanning around my old haunts. Maybe visit my sister-in-law's farm. I'd like to do a bit of fishing in New Zealand. And perhaps I could stroll about in Hong Kong and take a spin on the harbour. Oh! I haven't been to Cuba yet. I need to go there. I'd have a week in Bali, then go to India to listen to the music and eat some great food. That would be it. Perhaps I'd like to go to the Greek islands – I've got some airy-fairy ideas of what they would be like. Probably heavy on the tropical nikki-nakky-noo, though. Russia? It's a policy of mine to go to places you can get back from. North Korea? Fuck that. Tibet might be in that category. And I think I've missed China. People say they want to go to space. That's not my cup of tea. I'd be too scared to have a good time. I'd keep thinking: 'If this engine stops, we're fucked.'

The grand, intrepid journey might be a thing of the past for me now but, as a true Rambling Man at heart, I don't really need to do anything big. I can just go for a stroll to the end of my street and see the Muscovy ducks that were inherited by a neighbour. But she doesn't have a pond, so they waddle up and down the road and force motorists to stop and wait for them to cross. They're a lovely, shiny lavender colour in the light. I can ramble to the Cuban coffee shop and buy myself a cortadito . . .

but only if I want to stay awake for a fortnight. Nowadays, jumping on a moving freight train is not exactly my thing – and it never has been – but in my dreams and fantasies I'm a boxcar-jumping hobo; I stride out into the world with my banjo slung over my shoulder, meet people, sing songs, and have as many brilliant adventures as I please.

I suppose the Rambling Man's core philosophy is similar to that of the Buddhist, since he too avoids being overly tied to people, places or possessions and seeks the 'Way of Non-Attachment', aka freedom on the open road. Perhaps a Rambling Man in Eastern cultures would become a Buddhist monk, thereby legalising and legitimising his lifestyle and guaranteeing not only free food, but reverence as well. 'Billy, surely that's a bit of a leap?' I hear you say. You may be right, and I'm fucked if I can explain my own motivation; I just know I've always been drawn to the Rambling Man life and, even when I couldn't roam about the world, the sensibility still lived inside me. In fact, until some bastard invented mobile phones, I could roam free only minutes from my home and no one – including my wife and kids – ever knew where I was, what I was up to, or when I was coming home. It was glorious to be able to escape that way. The world of GPS, street cameras and tracking devices has dealt a crippling blow to the Rambling Man's entire existence.

But never mind. You can safely harbour your Rambling Man soul in almost any circumstance. In writing this book I've had the joy of returning to my favourite haunts without taking a single step – reliving many of my favourite journeys in my head. Maybe you too are a Rambling Man at heart . . . or perhaps you're one of the lucky people who can actually travel the highways and byways. Either way, it's a joy to be tramping merrily along your way, looking forward to a future journey, or simply imbued with longing for one you'll never take. Here's my best advice: sing as you go. And next time you're watching a football match – either live or on TV – and you happen to hear a crowd singing, tweak the words to a Rambling Man's anthem:

Walk on, walk on, with hope in your heart,
And you'll ever walk alone
You'll ever walk aaaaloone!

ACKNOWLEDGEMENTS

I consider myself a lucky man because, throughout the entire process of writing this book, I've had the brilliant support, advice and expertise of Nick Davies, the Managing Director of John Murray Press. In fact, it was Nick who launched me into authoring my own books, starting with *Tall Tales and Wee Stories*, then tackling my first autobiography, *Windswept & Interesting*. I'm very grateful to him – and to the entire talented and dedicated team, including:

Charlotte Robathan
Lauren Howard
Amanda Jones
Al Oliver
Kate Brunt
Sarah Arratoon
Alice Graham
Alice Herbert
Megan Schaffer
Drew Hunt
Kerri Logan

Rich Peters

Ellie Wheeldon

Dominic Gribben

Mel Winder

Suzy Maddox

Tans Mackenzie-Cooke

Lillian Kovats

Madison Garratt

Eliza Thompson

Katrina Collett

Juliet Brightmore

Jacqui Lewis

Kirsty Howarth

Linda Carroll

And everyone at John Murray Press

CREDITS

PICTURES

TEXT